Games with a Purpose (GWAPs)

FOCUS SERIES

Series Editors
Joseph Mariani and Patrick Paroubek

Games with a Purpose (GWAPs)

Mathieu Lafourcade
Alain Joubert
Nathalie Le Brun

iSTE

WILEY

First published 2015 in Great Britain and the United States by ISTE Ltd and John Wiley & Sons, Inc.

ISTE Ltd
27-37 St George's Road
London SW19 4EU
UK

www.iste.co.uk

John Wiley & Sons, Inc.
111 River Street
Hoboken, NJ 07030
USA

www.wiley.com

Library of Congress Control Number: 2015940033

British Library Cataloguing-in-Publication Data
A CIP record for this book is available from the British Library
ISSN 2051-2481 (Print)
ISSN 2051-249X (Online)
ISBN 978-1-84821-803-1

Contents

List of Figures

It should be noted that all figures are also available in color at www.iste.co.uk/lafourcade/games.zip.

Introduction

Since the 15th Century, the oxymoron *serio ludere* has suggested the idea of combining the concepts of games and *serious subjects*. Humanist literature would extensively use this concept to denounce all kinds of social problems. Then, until the development of computing, the armies of the world would exploit this concept through *war games*, which are ludic simulations for developing new tactics and training officers[1]. The modern concept of the *serious game* dates back to 2001–2002, with the video game AMERICA'S ARMY[2], developed for the US military to simulate training exercises. But almost simultaneously, the notion of the *serious game* would be popularized by its application to the educational world. Today, the title *serious game* refers to such diversity in terms of support, concepts, intentions, approaches and target audiences that it is difficult, if not impossible, to confine this concept into an exhaustive definition. One of the least restrictive definitions refers to *a computer application that combines serious intent, education-oriented, and that is informative, communicative, marketing, ideological, or for training, with recreational parameters issued from video games or computer simulations.*

Karen Chabriac, in her review of all the attempts of definition of serious games[3], concludes that the most synthetic is that of [MIC 06]: *any type of game whose purpose is other than mere entertainment.*

Due to the diversity of themes, objectives and approaches, there are numerous attempts regarding the classification of "serious games", depending on their purpose (advertising, recreational–educational, military, simulation, prevention, training, rehabilitation, etc.), or even depending on the *serious* function associated with the

1 http://fr.wikipedia.org/wiki/Jeu_sérieux/.

2 http://en.wikipedia.org/wiki/America' s_Army.

3 http://www.cndp.fr/savoirscdi/cdi-outil-pedagogique/apprentissage-et-construction-des-savoirs/education-et-pedagogie-reflexion/les-jeux-serieux/les-jeux-serieux-des-elements-de-definition/.

ludic basis (informative broadcast, educational, persuasive, military, etc.), provide training to improve the physical or cognitive potential of players, promote the exchange of data between players (or between the designer of the game and players), and/or the associated market sector (the type of public targeted). There is even an interactive site dedicated to the research of serious games based on several simultaneous criteria, divided into three categories: *intention*, *market* and *public*[4].

The general idea is that there are only two broad categories of games: *ludic games* designed for entertainment and *serious games* in which the player learns. In fact, in parallel to this concept, a third category of games has been emerging for slightly more than a decade, the Games With A Purpose (*GWAPs*), in which it is the player who teaches something to the machine. The concept of GWAPs is based on the idea of harnessing human skills for purposes of research and/or data production, whether destined to *support programs to progress in their understanding of the world* or more simply to use home computers to increase the computing power at the service of a research project. In either case, the ludic component is essential to motivate the public. The applications are numerous and the sector is growing. The concern is to exploit the *available brain time* (available and not only willing, but enthusiast if possible) to perform tasks that machines are (still) unable to do. A non-negligible or even crucial aspect is that this type of games, directed toward the production of all kinds of data, makes it possible to, therefore, use the creativity, the imagination, the knowledge and the know-how of hundreds or even thousands of users at a lower cost. It should be noted that unlike the AMT system[5] (resource collection tool that uses crowdsourcing and offers a derisory remuneration as well as conditions not complying with the French labor law) the principle of GWAP does not raise any ethical problem, as long as it remains free and does not offer prizes that look like disguised salaries [SAG 11].

Originally, at the formalization of the concept of GWAP, there were CAPTCHAs, which were invented by Luis Von Ahn [AHN 06a], an American academic: these are small tests based on the deciphering and the input of a sequence of characters, which are employed to differentiate a human from a computer on the Internet, and thus to prevent spam, phishing or any other malicious activity by *automatic* means. Luis Von Ahn realized that the 10 s spent by a human to decipher a CAPTCHA (therefore to do something which a computer does not know how to do) could be usefully employed. He then created RECAPTCHA: from now on, when a captcha is decrypted, not only does it identify the user as a human being, but it also helps to digitize books by deciphering sequences of characters that the optical character recognition (OCR) is unable to decipher. The principle of the GWAP was born, and will be illustrated by ESP GAME [AHN 04]: the father of CAPTCHAs invented a game that consists of presenting the same image to two players who will score points

4 http://serious.gameclassification.com/FR/.

5 Amazon Mechanical Turk

and progress as soon as they suggest the same keywords to define it. The interest is naturally to make searching for images by keywords in a search engine more powerful, accurate, fast and relevant.

The use of GWAPs, either to collect data, annotate images or documents, or to solicit the public for solving major scientific problems, is currently expanding and it concerns all areas. Nonetheless, it gives rise to the most high-profile experiences and results in the life sciences and medical fields. The concept of *citizen science* reflects not only a change in the way that scientific issues are perceived by the public, but also a willingness to take science out of laboratories and researchers out of their ivory towers. By making the challenges of research accessible and understandable by ordinary people, science is demystified and desecrated. For researchers, the ludo-collaborative approach is a powerful way to involve the public and to mobilize its support and its empathy, while soliciting and evaluating its non-specialist outside perspective. It should be noted that, according to Luis Von Ahn and during the years 2005–2010, almost 10^{10} (i.e. 10 billion) hours have been spent annually by individuals playing on the Internet (which globally corresponds to an average slightly higher than 1 h per person and per year). Why not try to deviate from it, at least a tiny portion, for useful games allowing the acquisition of resources?

GWAPs are thus numerous in multiple disciplines, and it is now obvious after a careful overall review that the effort of giving a really recreational dimension to a useful task is very uneven. As a result, games where the interest is all the more stimulated by a real challenge, by emulation among players, by a motivating classification system and, above all, by a real interest of the underlying task are found less often than games stimulated by the idea of helping science and/or of doing good work. Frequently, various parameters are borrowed from the universe of games (design, avatar, sounds, etc.) to give a ludic polishing to a monotonous and repetitive task. It is clear that in many cases game designers primarily rely on the players' *civic zeal* by means of the excitement caused by *participative* science and the extremely rewarding feeling to achieve something useful in a field which remains prestigious in the eyes of the general public.

With regard to GWAPs, and since they are all based on the principle of crowdsourcing, large discrepancies can be observed according to the nature and the magnitude of the task, the target audience, the field of research, the skills required on the part of the players, the ludic and/or the educational dimension, and the manner in which the data generated by the game are processed. The choice of criteria for establishing a classification is, therefore, difficult; however, the classification of [GOO 13], which is based on the nature, the extent and the complexity of the task, can be considered as interesting. In fact, these authors make the distinction between *microtasks* and *macrotasks* in systems using crowdsourcing for scientific purposes. Microtasks are tasks that can be solved within seconds by anyone who is able to read some simple instructions, and macrotasks, on the contrary, concern complex problems that are resistant to the qualified experts of the institutional research. The first category of tasks require a large number of people who will process a huge volume of data in a short period time, and whose contributions, (strongly) redundant between players, will be aggregated in order to provide data of a quality as good as experts' annotations. For the second category, resorting to crowdsourcing makes possible to detect the few talented people within a large population of potential candidates with very heterogeneous skills. Through the provided interactive environment, they will not only demonstrate but also develop the inventiveness, the curiosity and the creativity necessary to meet the challenge and allow real scientific progress.

It should be noted that among GWAPs, a huge majority concern microtasks.

It should also be noted that games with a large audience whose spectacular results are subject to a wide media coverage, and which are generally macrotasks (FOLDIT[6], EYEWIRE[7]), all have a dedicated website while GWAPs, with a more modest ambition, instead related to microtasks, are often proposed through portals. The most famous and the oldest is ZOONIVERSE[8].

At the origin of the ZOONIVERSE portal, there is GALAXYZOO[9], an online astronomical project which somewhat symbolizes the beginnings of the type of science known as *citizen*. GALAXYZOO is a scientific project based on volunteers, destined to characterize galaxies from pictures, and the success was such (in 2007, 85,000 contributors in five months, according to Wikipedia[10]) that the idea to secure the cooperation of the general public has inspired many other programs. Currently, the ZOONIVERSE portal is a huge international platform, which brings together a large number of collaborative projects among which some have borrowed attributes from games to present a more attractive appearance. After registering (more than one million people have registered around the world), it is possible to collaborate in all the projects of the portal, which are classified by main themes (space, weather, nature, humanities and biology). Within each theme, various crowdsourcing activities are proposed, almost all dedicated to the processing of a large volume of data. These activities can either be identifying fauna and flora on pictures of seabeds (SEAFLOOR EXPLORER), or wild animals which pass through the field of a camera installed in a natural reserve (SNAPSHOT SERENGETI). It can also involve the deciphering of handwritten labels of animal or botanical specimens kept in museums of natural history (NOTES FROM NATURE), as well as many other useful tasks and for which automated processing is excluded. All of these activities fall within the scope of the

6 http://fold.it/portal/.

7 http://eyewire.org/.

8 https://www.zooniverse.org/.

9 http://www.galaxyzoo.org/.

10 http://fr.wikipedia.org/wiki/GalaxyZoo.

voluntary participation rather than that of gaming. This is true despite the presence of ludic elements such as a *score*, or the creation of a *collection* that brings together all the processed specimens, or even the granting of *ranks* sanctioning the relevance of the work done (promotion from *cadet* to *lieutenant* then to *captain* by deciphering and transcribing weather information from logbooks of American ships dating from the mid-19th Century in OLD WEATHER, to contribute to the study of the evolution of climates). This is why, with the exception of WORM WATCH LAB, we did not review these games, whose ludic side is nothing more than a quite unconvincing vague layer, the motivation being instead stimulated by the rewarding idea of participating in *real science* (*Real Science Online*).

In summary, we can say that GWAPs are games, useful games, but essentially games. They are useful for the community; these are games for players. The designers of GWAPs must not ignore this duality, incurring the risk of either obtaining very few resources (if the players get tired too quickly) or of obtaining low-quality resources (if the ludic aspect is developed at the expense of the utility aspect). We will return to the characteristics and limitations of GWAPs in more detail in Chapter 5, as well as in the Conclusion of this book. GWAPs are thus games with a purpose, even if all games have a purpose, mainly to entertain or to teach. We should rather talk about *games for the purpose of resource acquisition* or in some cases *games for the purpose of problem-solving*.

In this book, we present and analyze a number of GWAPs according to a thematic organization. Of course, this list is far from being exhaustive: during the drafting and the printing of this book, new GWAPs will have appeared (and some will have disappeared). Chapter 1 focuses on GWAPs related to the field of biology in a broad sense. Chapter 2 is specifically focused on games with a medical purpose (but not those with a therapeutic objective, which are not GWAPs). Chapter 3 describes GWAPs concerning language, linguistics and natural language processing. Chapter 4 deals with GWAPs that do not fall within any of the above categories. Chapter 5 presents and analyzes in detail the JEUXDEMOTS project whose objective is to build a large lexical knowledge database using games. Finally, the Conclusion draws upon some lessons from the JEUXDEMOTS experiment, and from the overview of other projects, to attempt to define what criteria are significant and must be privileged in the design of a GWAPs in order to transform it into an effective data acquisition tool.

1

Biological Games

In this chapter, we present the main GWAPs currently available in biology and biochemistry.

In the highly specialized field of molecular biology, the results of FOLDIT[1] have been so positive (three articles published in the journal *Nature* – [COO 10, MAR 12] and [HAN 10] – and another in the Proceedings of the National Academy of Sciences, or PNAS [KHA 11]) and media coverage so important (a large number of articles in the non-specialized and popular science press) that the concept of GWAPs has quickly spread to biology. Many researchers try, with varying degrees of success, to use an entertaining-contributory approach to solve complex problems.

Zoran Popovic, one of the creators of the game, and director of the *Center for Game Science* at the University of Washington, believes that his laboratory is responsible for the emergence of a new path of scientific discovery. This path takes advantage of the creative capabilities of a wide population, once it is possible to involve it in research mechanisms through an effective initiation.

Adam Gazzaley, professor of neurology, physiology and psychiatry at the University of California and creator of NEURORACER (a game for improving the cognitive abilities of older people), being more restrained, thinks that the field of serious games is still in its infancy and that there should be a selection between that which claims to be based on science and what has actually been validated by scientific methods. According to him, approaches and tools should be multiplied and the results should be verified before making strong recommendations.

1 http://fold.it/portal/.

1.1. Foldit

Type of task – Macrotask.

Preliminary remark: FOLDIT is certainly the ancestor of biological GWAPs dedicated to the solution of complex problems. One of its designers, Adrien Treuille, has also participated in the creation of ETERNA[2] (see section 1.2), and the FOLDIT team is also at the origin of the *Center for Game Science*, an organization of the University of Washington dedicated to the universe of gaming for scientific purposes. The *Center for Game Science* hosted a very recent biological GWAP: NANOCRAFTER[2] (see section 1.3), a construction game of molecular nanostructures.

Launch – May 2008.

Audience/popularity – 57,000 players 2 years after launching.

Goal – Create predictive models of three-dimensional (3D) structures of proteins from their amino acid composition.

Significance of crowdsourcing for the problem to be solved – Predicting the 3D structure of a protein by its amino acid composition is a fundamental issue: a large number of diseases are caused by mutations, which by affecting the 3D structure of a protein, alter their normal operation. If the *normal* and the stable structure of a protein can be predicted, it can be understood how a mutation occurs at the level of the spatial conformation, and appropriate adapted therapies can be developed (vaccines, proteins for therapeutic use, etc.).

At the origin of FOLDIT, there is the distributed computation software program Rosetta@home: this is a program that determines the possible conformations that may adopt a protein by calculating the probability of interactions between segments of polypeptide chains according to the amino acids that compose them; interaction chains being all the more probable that the level of energy required for them to take place is minimal. Therefore, the most probable 3D structure is the one that will require the least amount of energy to develop. However, for a given amino acid sequence, there exist so many possible spatial conformations that simulation with algorithms is an extremely long and complex process. Rosetta@home uses home personal computers and runs when they are hardly used or not, thus increasing the computing power available to the project known as *distributed*. FOLDIT has been designed as an interactive and recreational extension of Rosetta@home.

According to David Baker [HAN 10], one of the researchers at the origin of the project, FOLDIT players are not required to have simple skills of visual recognition,

2 http://nanocrafter.org/.

of classification; they are asked to utilize their logic, intuition, ability to see in 3D, initiative and creativity in solving complex optimization problems. Baker says to use three skills of the human brain, three typically and exclusively human talents:

– the ability to know how to take a risk in the short term for a long-term advantage and *vice versa*;

– the ability to see in 3D;

– the ability to realize that a *dead end* has been found and to stop just in time (that is before too much time has been wasted).

Zoran Popovic, one of the main stakeholders of the project (of its software/computing aspect), claims that for the time being, the human brain remains better than the machine as soon as the *creativity* parameter becomes essential and unavoidable. In addition, the game dimension creates the ideal conditions for the emergence of a collective expertise: the designers of FOLDIT have observed that not only do the players improve over time, but also that they organize themselves to add their respective competences. Finally, the game is designed in such a way that the players' findings are reinjected, recycled in order to be integrated in the manipulation mechanisms of the virtual protein, such that new players become increasingly faster experts.

Thus, a typical case (illustrating the first of the three abilities mentioned above) where humans are performing better than machines is that where a structure must be profoundly reshaped to optimize its configuration, that is to reach a stable form with a minimal energy cost. This implies too many sacrifices in terms of computation time for the machine, since this requires, as a first step, disorganizing an existing layout.

The most passionate gamers showed such skill in manipulating the protein chains, that designers have rapidly updated the game so as to provide the ability to create *ex-nihilo* proteins (potential applications: finding molecules likely to catalyze certain metabolic functions, manufacturing vaccine candidates, designing molecules that meet specific biological constraints for biotechnological or medical applications, etc.).

Moreover, David Baker uses the players' findings, that is to say specific folding strategies, to automate them in order to improve the prediction algorithms of protein structures. The community of players is regarded as a scientific partner on its own, and is associated as a coauthor in scientific publications [KHA 11].

Qualities/required human skills – Intuition, logic, spatial vision ("humans have 3D understanding that computers just cannot handle yet, and computers have the number crunching capabilities that humans cannot possibly compete with"), deduction and initiative skills, curiosity and creativity.

Necessary learning/game accessibility – An initiation step, necessary to get familiarized with the concept and the tools, is provided by means of a tutorial. It consists of a succession of 31 *puzzles* which reveal through step-by-step explanations all the mechanisms and modes of actions available, to act on the proposed 3D structure in order to bring it into an accurate conformation. It is possible to confront the real *puzzles* and to compete with other players before solving the 31 steps of the tutorial, but the relative complexity of the mechanisms to control depending on the various possible case studies makes it rather random. The manipulation of the structure modeled in 3D proves to be quite difficult and confusing, and it happens frequently that the solution of a *puzzle* depends largely on a lucky guess rather than on the real mastery of the suggested tools. This corresponds exactly to the *hardgame* configuration described by Good and Su [GOO 13].

Player's education – During the learning phase with the tutorial, the player learns the fundamental biochemical rules that govern the folding of protein chains: he/she must minimize the empty spaces within the molecules by promoting the creation of hydrogen bonds, manipulate the sidechains so that the hydrophobic parts are oriented inward and avoid excessive proximity between the atoms. The modeling of the molecule allows the visualization of these parameters and the mechanisms that can act upon it. Without becoming experts biochemists, regular and passionate players eventually end up acquiring an extensive expertise with regard to the spatial conformation of the protein, and for some of them, the ability to intuitively understand what must be changed to improve the 3D structure. This justifies in the eyes of the developers the benefit of entrusting non-specialists with the design of potentially synthesizable virtual proteins.

Interactivity – It is highly developed: players do not interact through the actual game since each is alone with his/her protein structure to optimize, but the possibilities to communicate with other players and designers are multiple: an instant messaging window is available in the interface where the protein is manipulated, and the number of players connected at the same time on the same configuration can also been seen (several *puzzles* are simultaneously proposed). From the homepage of the game, where the latest news are updated almost daily, a blog is available, where designers detail the scientific progress, and the results are achieved through the activity of the players; a forum and a wiki can also be found.

Challenge/emulation – As a result of the learning phase, where one can practice alone and at his/her own pace on models of proteins whose structure is known, *puzzles* can be tackled, therefore, protein structures to be optimized, which are put into play for a fixed duration. A *puzzle* at the beginner level can be chosen, or a *puzzle* depending on its characteristics (*overall* (generalist), *prediction* (known amino acid composition, unknown 3D structure), *symmetry* (proteins consisting of multiple identical assets arranged along a symmetry axis). The player is then in real-time competition with all the other players who have chosen this *puzzle* and who strive to improve the structure. There are between 0 and 50 players simultaneously connected to the same *puzzle*, and things take a pretty exciting turn since the objective is to raise his/her score and thus to rise in the rankings, which can be followed in a small window embedded on the page. The points counter reacts to the slightest intervention on the structure, and an animation with a sound effect appears on the page as soon as the score exceeds that of another player (*rank up*), but also if one regresses in the ranking. The desired aim is obviously to stimulate the player so that he/she "hacks" his/her protein until he reaches the top of the rank. The time limitation spices up the challenge.

Design/feeling – The 3D modeling of the molecule is very successful, it can be manipulated and easily directed with the mouse and thus easily seen from multiple angles to identify problem areas (too much vacuum, atoms too close, hydrogen bond to favor, etc.). It is very reactive on the whole, each action generates an effect; fortunately, any action can be canceled, and the possibility of returning to the best configuration obtained is reassuring and allows all the initiatives: the best score and the corresponding conformation can be restored even if starting an attempt to improve that eventually worsens things (which is frequent…). The sound effect of the different actions is nice, the background music is not nagging and anyway, the interface is configurable with in particular the ability to turn off the sounds.

1.2. EteRNA

Type of task – Macrotask.

Launch – End of 2010.

Audience/popularity – More than 50,000 registrations, many articles in non-specialized and scientific press, one of which in the *Proceedings of the Academy of Sciences*, where players are associated as co-authors [LEE 14].

Goal – The purpose of ETERNA[3] is to virtually design complex ribonucleic acid (RNA) molecules which will be really synthesized, and tested to verify that they correctly adopt the 3D conformation theoretically predicted, and that they are biologically active. Given the key role that RNA plays in protein synthesis, and the role of proteins in living organisms, it is understood that the mastery of synthesis and RNA conformation is at the base of multiple medical, therapeutic and biotechnological applications.

3 http://eterna.cmu.edu/web/.

Benefits of crowdsourcing for the problem to be solved – Human brains prove to be more effective than computers in predicting the ways in which a molecule of a known nucleotide sequence can be folded. It should be noted that this assertion seems controversial: nondeterministic polynomial time (NP)-complexity does not seem proven for the design of RNA, as reported by [TCH 14]. (A problem is known as *NP-hard* when it cannot be solved within a reasonable time by an algorithmic method and that thereby justifies the use of the participatory approach).

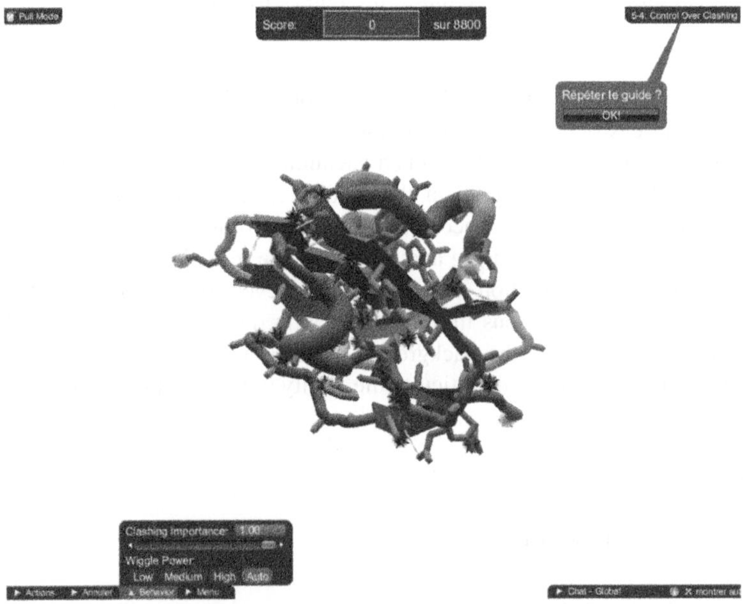

Figure 1.1. *Foldit. The game interface is a three-dimensional view of the protein being studied. The molecular components (hydrogen bonds, hydrophobic chains, etc.) can be manipulated with the mouse, and each action alters the spatial conformation of the protein*

Qualities/required human skills – Logic, deduction ability, initiative, curiosity, creativity, strategy, etc.

Necessary learning – The player is introduced to the rules of pairing of nucleotide bases by means of a tutorial consisting of a series of *puzzles* that he/she can solve through step-by-step explanations.

Player's education – Although the bases pairing rules are very simple, the parameters that determine the stability of the molecule can only be gradually apprehended, as different spatial configurations are found, and by reading what regular players publish, which eventually end up developing real expertise in this extremely specialized area of molecular biology.

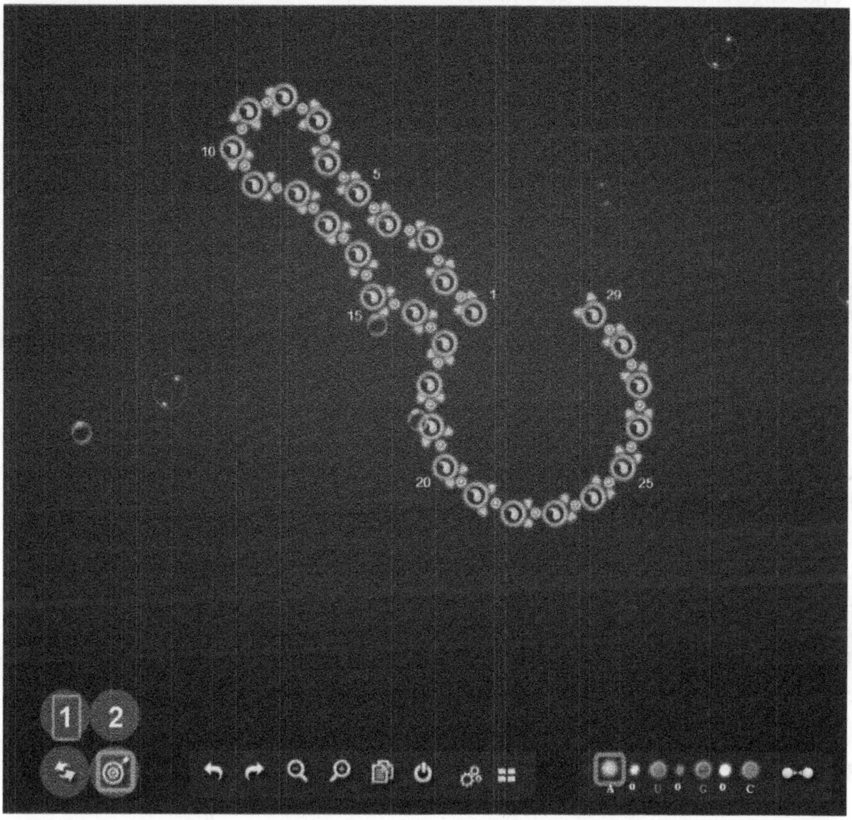

Figure 1.2. *EteRNA. Here, an RNA molecule in the spatial conformation that it must adopt if the proper sequence of nucleotides is performed, the one that will allow the establishment of bonds ensuring the stability of the structure. The circles in the molecule are to be replaced by the A, T, G and C bases chosen at the bottom right*

Interactivity – As in the case of FOLDIT, the interactivity between players is privileged: instant messaging in the game is very active and almost always shows between 40 and 50 players simultaneously connected, which seem to form a community of regulars who know each other and communicate. On the other hand, for each proposed *puzzle*, there is access to a comment space where players who wish to do so can express themselves, give clues or explain how they came to grips. Finally, the forum of ETERNA is also highly frequented and there is also a wiki, mainly fed by the community of players. This wiki contains a lot of strategies, tips and explanations to solve the *puzzles*.

Challenge/emulation/task gamification – Players are not in direct competition with each other; they solve *puzzles* to earn points, climb in the ranking and cross multiple *grades*: *Tutorial completed*, then *Puzzle master LV1*, *LV2*, etc. (depending on the number of solved puzzles) and then *Top Player LV1*, *LV2*, etc. (be part of the best *n* players). At the same time, by solving *puzzles points* are accumulated and the player successively becomes a *Lab Member* and then a *Puzzle Architect LV1, LV2 and LV3*. The ultimate grade is that of *Millionaire*, reserved for players having accumulated 1,000,000 points playing with ETERNA. On the other hand, conviviality and the interactions between players are also parameters of the ranking, and are strongly encouraged by *grades*: *comment* (having at least posted five comments), *follow players* (follow at least five players), *gather followers* (having at least five *followers*). Note: the multitude of *grades* and ranking parameters (i.e. the number of completed puzzles, position in the ranking, earned points, conviviality rewards) mean that even beginner players immediately see the *counters* progress (points, number of *puzzles*, rank, etc.), which is a form of stimulation and encouragement. Their profiles show them what they have *unblocked*, but also all the steps that remain to be done.

Design/feeling – The design is neat, the interface is clear, help in readily availabile and easy to access the colors are harmonious, etc. Animations and sounds give a rhythm to each action, the ergonomics are good and it is possible to zoom at will on the molecule. The deep blue lit by a halo and the bubbles that slowly rise give the illusion of a water environment which is relaxing, serene and zen.

1.3. Nanocrafter

Type of task – Macrotask.

NOTE.– NANOCRAFTER was launched in late April 2014 by the same team as FOLDIT, in the framework of the *Center for Game Science* of the University of Washington, with these words from Zoran Popovic:

> "After more than a year in the making, NanoCrafter is ready for the first public Beta release. Over the many years to come, we expect the game to evolve almost on a daily basis towards a vibrant community of creative DNA designers, hopefully producing many exciting novel designs. As with any Beta we expect many things need to be improved, we look forward to improving the game on a daily basis. I am announcing the beta release of the game at Games for Change in NYC today."[4]

4 http://nanocrafter.org/.

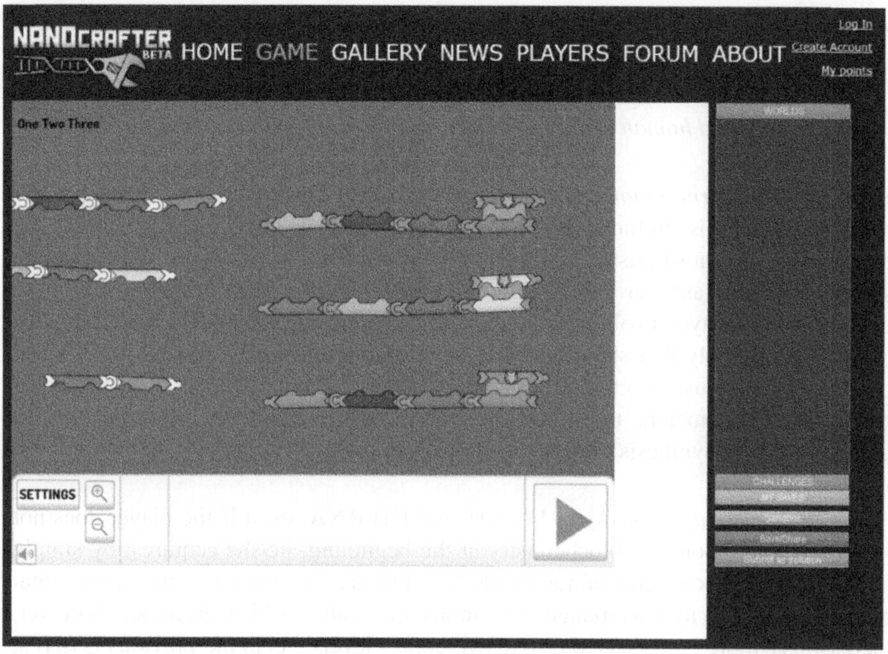

Figure 1.3. *Nanocrafter. In the configuration above, colored items represent fragments of single-stranded DNA that must be assembled respecting the color correspondences to form double-stranded structures*

Goal – NANOCRAFTER approaches the theme of the synthesis of desoxyribonucleic acid (DNA) fragments by using the pairing properties of nucleotides forming DNA, and especially the displacement mechanisms of strands that occur spontaneously when several single strands of compatible sequences are brought together. The objective, through the game, is to think about the construction of nanostuctures whose potential biotechnological applications raise a growing interest in structural biology and especially in nanomedicine.

Benefits of crowdsourcing for the problem to be solved – NANOCRAFTER is directly inspired by what works in FOLDIT and ETERNA; these are, therefore, the same human skills that are looked for through the proposed tasks. The main difference is the emphasis placed on creativity since the purpose is to invent molecular assemblies in order to create structures that do not exist in nature, on the basis of the building blocks of life (DNA) and from what is known of their constraints and properties of assembly. The field of biological synthesis is still in its infancy, but applications in health matters are potentially significant, and according to

Zoran Popovic, it is the privileged domain to express the advance that the human brain still has when compared to the machine.

Quality/desired human skills – Creativity, curiosity, initiative spirit and logic.

Necessary learning/game accessibility – As in other games of this kind, the beginner player is initiated to the mechanisms that govern the assembly of nucleotides, and must pass through the different stages of initiation by performing simple building tasks at the beginning, and which become progressively more complex. The player evolves through *worlds* which correspond to techniques of molecular assembly that should be mastered, each world being divided into several thresholds that must be crossed to unlock the next. Thus, there is a minimum of biochemical parameters to know and whose utilization must be controlled to undertake the biosynthesis of biological nanostructures.

Player's education – As in FOLDIT and ETERNA, even if the player does not know anything about the biochemistry at the beginning, he/she acquires, by simply playing, the basic concepts of nucleotide base-pairing, and the minimum of rules that must be known to manipulate the molecule, rules which become, however, increasingly more complicated as one overcomes levels. As in the previous GWAPs, this might be sufficient for him/her, and he/she does not need to know more in order to play. The game is still too recent, but it is easily imaginable that, as for ETERNA and FOLDIT, the more *hooked* players will quickly acquire expertise nurtured by their practice on the one hand, emulation on the other hand, and the assiduous attendance of discussion spaces of the game (forum and news), where the design team, with a strong presence, regularly communicates information and developments with a more or less pronounced popularization effort. The impression is given that non-scientists, who in appropriating the codes, the vocabulary and discussing issues between them, are gradually acquiring a highly specialized culture, focused on the problems raised by the game. Researchers, for their part, are doing the minimum about popularization in order for the general public to understand what to do (the game has a very detailed help topic available). The reconciliation between researchers and the general public is more pronounced in the sense where the general public acquires specialist knowledge than in that where the specialists try to reach the general public…

Interactivity – The game was recently launched, but like its predecessors, the possibilities of interaction and communication are highly developed: players have access to an instant messaging window embedded inside the game window, to a forum, and a wiki will probably complete the whole.

Challenge/emulation/gamification – The game is too recent to tell, but since a team that wins cannot be changed, the mechanisms that have made the success of

FOLDIT are more or less reused. Outside a solitary progression through *worlds* where levels are gradually unlocked by discovering new ways to act on building blocks, *challenges* are proposed, which are objectives to be fulfilled with constraints to respect, for example *achieve a double-stranded DNA fragment using the parts below*. When it is estimated that the objective has been achieved, the found solution is submitted, which will collect points if it is validated. For the moment, nothing explains clearly how and by whom the proposed solutions are validated, nor the calculation method of the points, probably because the game is still new. Therefore, rather scarce emulation and competition (it is unclear whether several players attempt simultaneously the same challenge, and it is not known whether the players vote for the various proposed designs), the bulk of the interest comes from the *puzzle* aspect, and from the suspense related to the validation or invalidation of the proposed structure. It should be noted that in the levels of the different *worlds*, the answer (valid or invalid) is given at the submission, with as an extra, the reporting of problematic assemblies in the case of failure. Conversely, in the *challenges*, there is no immediate news about the solution that was proposed.

Design/feeling – The interface is nice, clear and colored; the building blocks move with the mouse with fluidity; the manipulation is intuitive; assembly rules are accessible at any time by means of a button in the manipulation window; a structure can be put into the trash and one can start over; in case of failure, there is the possibility of choosing between restarting from scratch or recalling the last configuration achieved and modifying it to try to correct what is wrong.

1.4. Phylo

Type of task – Macrotask.

Generalities – The creator and his team emphasize the fact that PHYLO[5] is an open-free project and define their game as a new tool to support the analysis of the genome available to the international scientific community. A link *contribute to Phylo* invites users to collaborate by improving the code at the base of the game, or by translating the game into different languages (the displayed ambition is that PHYLO be translated in all possible languages; it is currently in 11). Furthermore, since October 2013, researchers around the world have been able upload their sequence alignments on PHYLO and submit them to the ability of the players to improve them. This is done through an interface called OPEN-PHYLO, which requires a separate registration and through which researchers not only upload their own sequences, which need to have been prealigned, but also generate and organize the *puzzles* on the portions of their choice [KWA 13].

5 http://phylo.cs.mcgill.ca/.

Jerome Waldispülh, bringing forward the *open-source* status of his project, raises the problem of the status of the data generated by crowdsourcing, deploring the fact that most of the time they are not free. He also points out that the first crowdsourcing projects have, because of the novelty of the concept, received considerable media coverage, which has greatly facilitated their dissemination. Support by strong media, with very wide audiences, is the *sine qua non* condition in order for this type of project to meet the audience it deserves. However, the number of projects known as *citizen science* is constantly increasing. The effect of novelty, therefore, fades away, and the trivialization of the concept will make public adhesion increasingly difficult and random. Even worse, according to Hand [HAN 10], journalist at the journal *Nature*, if the concept of GWAP is disproportionately extended, at some given point the general public will no longer be amazed and flattered to be asked to contribute for science. This is a long way from saying that people are instead likely to feel exploited and manipulated, and that the day will come when scientists will have to pay the volunteers to ensure the support of the public. There is a step that is still far from being taken, but Michael Kearns, computer scientist at the University of Pennsylvania, dreads the development of economic activity around the concept of *citizen science* [HAN 10].

For the creator of PHYLO, web portals such as ZOONIVERSE[6], CITIZEN SCIENCE CENTER[7], or SCISTARTER[8] can to some extent compensate for the problem of the lack of media coverage which new projects will have to face as the concept of GWAP will become commonplace. In order to not dilute the potential power of the crowd in a multitude of projects, he suggests that the research teams come together and use the same game interface when they have the same type of problem to solve, rather than to each create their small GWAP on their own. This is why he puts PHYLO at the disposal of the international scientific community. This is what he explains in the blog of the game by presenting the OPEN-PHYLO interface. The existence of a game such as FRAXINUS[9] (see section 1.5), more or less modeled on PHYLO, is the perfect counterexample to what Waldispülh advocates.

Launch – Launched in October 2010 by Jerome Waldispühl and his team at the McGill University in Montreal, the game has been widely commented in scientific [STR 12] and general media, and as FOLDIT and ETERNA, it has been the subject of a scientific publication to which the players have been associated as co-authors [KAW 12].

Audience – More than 300,000 people since the launch according to the online scientific media *Biomedical Computation Review* .

6 https://www.zooniverse.org/.

7 http://www.citizensciencecenter.com/.

8 http://scistarter.com/.

9 https://apps.facebook.com/fraxinusgame/.

Goal – Researchers (biochemists, geneticists and evolutionists) need to compare sequences of macromolecules (notably DNA, as well as RNA, proteins, etc.) originating from different species, in order to detect similar areas. One of the applications is to deduct from the rate of similarity between the sequences, valuable information about the relationships between species. Another is also to identify, through alignment zones, important functional sites, or even to highlight the mutations, or to understand the origin of genetic diseases. Multiple sequence alignments are performed by algorithms, and typically represent a problem *NP-hard* (that is it cannot be solved within a reasonable time by a computer given the complexity and the cost in terms of computation time, without even mentioning the financial cost), which justifies the entertaining-contributive approach. The purpose is therefore to improve the sequence alignments obtained through computation algorithms by the players. The game presents itself as a puzzle where the objective is to find the best vertical alignment between successions of colored blocks which represent the four DNA bases. PHYLO thus allows the general public to do what was usually done manually by researchers: reorganize the bits of sequences to try to improve the alignment provided by software.

Benefits of crowdsourcing for the problem to be solved – Finding the best alignment compromise for a large number of sequences is thus a *NP-hard* problem. Realizing that performing this task manually did not require any knowledge in genetics but simply a good sense of observation, and abilities of reasoning and logic, Jerome Waldispühl had the idea to transform the problem of alignment of multiple sequences into a game of manipulation of colored geometric forms. In addition, he has found that in more than 40% of cases, players are able to improve the alignment created by computational algorithms. "Unlike an algorithm, human beings are able to circumvent the rules of the game to obtain the alignment" he claims.

Qualities/desired human skills – Sense of observation and deduction, reasoning and logic.

Necessary learning/game accessibility – PHYLO is designed to be as simple and quickly accessible as the famous Tetris. The simplicity of the task at hand is such that no learning is required, and that the game is immediately accessible: everyone, at any age, is capable of sliding colored squares along superimposed lines, with the objective of creating columns of the same color. Furthermore, since these sequences (lines) are not identical, it is necessary to find the best compromise between the inevitable color differences and *holes* in the line. A tutorial is nevertheless proposed, but mainly to explain how the points are calculated and what strategies should be preferred to improve the score.

Player's education – In games such as FOLDIT, ETERNA and NANOCRAFTER, the player is introduced to the rules of base pairing, acquires knowledge of the compared solidity of atomic bonds, must take account of notions such as concentration, and know and use various biochemical properties. Unlike the above games, PHYLO does not require any genetic or biochemical knowledge and the actual game does not provide any scientific information. This is an option that was chosen and that aims at simplifying the problem to solve into a game as simple and popular as possible, into a colorful puzzle that looks like any little online game that can partially and quickly be played in order to relax. Science becomes almost invisible... When the game starts, the subtitle *a puzzle to contribute for the research on genetic diseases* as well as the possibility to choose a *puzzle associated with a disease* are still there to encourage people to play by striking the sympathetic chord of good conscience. Naturally, the curious player can easily find what PHYLO is for, and scientific information is available in a blog and in the *news* on the homepage of the game[4], which are essentially links to press articles. As Jerome Waldispühl highlights, "The player is distracted but this is not lost time, and at the end of the match he is told about which gene related to what pathology he has allowed to move forward".

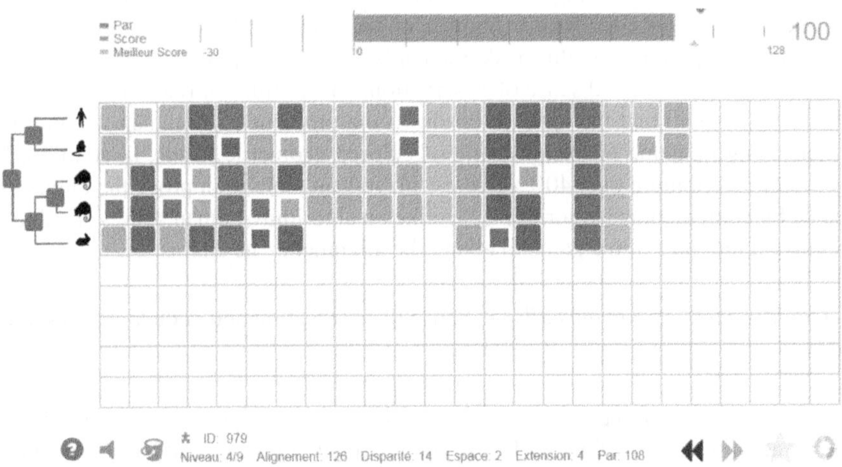

Figure 1.4. *Phylo. The purpose is, by moving laterally the color squares, to find the best alignment compromise possible between the five presented fragments of sequences. Above the sequences, the colored bar illustrates the progression of the score, the small top triangle being the score achieved by the computer, which must be met and exceed to move to the next level*

Interactivity – This is almost not developed. PHYLO is a solitary game, such as *Solitaire*, etc. The player tries to do better than the computer, and the only reference to other players is the ranking. There is a forum, but it is not really active. On the home page, a window shows the tweets related to the game in real time.

Challenge/emulation – The main attraction of the game is its immediate accessibility and the assurance of very easily scoring points at first, sufficiently enough to make the player *hold on* when it becomes more difficult. Thus, a *puzzle* is composed of several *levels*, each level adding sequences to align. Matching the score of the computer (the *Par*) is sufficient to access the next level, and it is rather easy in the early stages. When the player manages to complete the *puzzle*, he/she accesses the statistics and the information about this *puzzle*, which are an element of motivation, since he/she reads his/her score, the mean score, as well as the best score obtained for this *puzzle* and the name of the holder of the *record*... He/she then has the possibility to replay the *puzzle* to try to match or exceed the *record*. Three classifications enables the player to access his/her progress: monthly, daily and overall; the daily ranking is very motivating because the player can quickly progress. Achieving a *record* is also flattering and encourages users to replay. The goal of Jerome Waldispühl was to create a similar game to Tetris, whereby any person could play when he/she has a little spare time, and it seems to be totally successful: the rules are simple enough such that one game suffices to assimilate them, and a game is fast enough so that the time invested is not an obstacle.

Design/feeling – The design of the game is pleasant, the interface is customizable (the color of the plate and the constitutive bases of the DNA can be modified, action sounds and background music can be enabled/disabled) and the language can be chosen. The manipulation of blocks is fluid and the ergonomics is satisfactory. The result of the actions can immediately be seen through a highly visible progress bar, as well as the score to win (the *Par*). However, it is regrettable that the alignment achieved by the computer is not revealed when the player is unable to match the *Par* and when he/she decides to abandon it. This would be instructive...

1.5. Fraxinus

Type of task – Macrotask.

Goal/launch – This game[10] is a game of sequence alignment in all respects comparable to PHYLO, launched in December 2012 by the Sainsbury Laboratory, an English research laboratory specialized in plant pathology. Unlike PHYLO, it focuses exclusively on research concerning the ash tree disease caused by the *Chalara* fungus and is accessible only to members of the social network Facebook.

10 https://apps.facebook.com/fraxinusgame/.

The players improve sequence alignments issued from the pathogen agent and the affected tree to provide information that could help researchers understand how the fungus kills trees. Especially, how and why some ash trees are resistant is of paramount importance to develop a strategy to fight and eradicate the disease that is decimating ash trees of the United Kingdom.

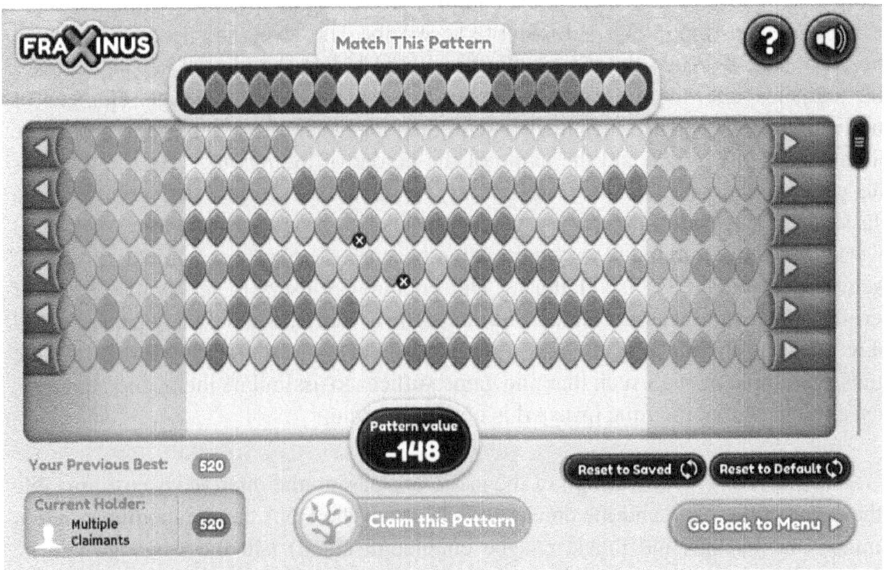

Figure 1.5. *Fraxinus: sequences to be aligned. The purpose is to find the best way to vertically align the colors of the blocks constituting six sequences. They must be aligned between themselves AND with the visible reference sequence at the top. This is achieved by pushing laterally the sequences using the left and right arrow keys*

Benefits of crowdsourcing for the problem to be solved/required human qualifications/game accessibility – Exactly the same as for PHYLO.

Player's education – As for PHYLO, and for the same reasons, it is not necessary to have knowledge of genetics and biochemistry to be effective with FRAXINUS. However, enough detailed information about the disease, the devastation caused to trees and the research strategies to eradicate this scourge is provided at the beginning of the game. This is used to suggest adhesion and motivation by striking the sensitive chords of patriotism and the keen interest for the conservation of nature: "Help us, otherwise 90 to 95% of ash trees in the United Kingdom will die..." At the end of each successfully achieved *puzzle*, the words "You are helping to save the ash tree" appear as a reminder.

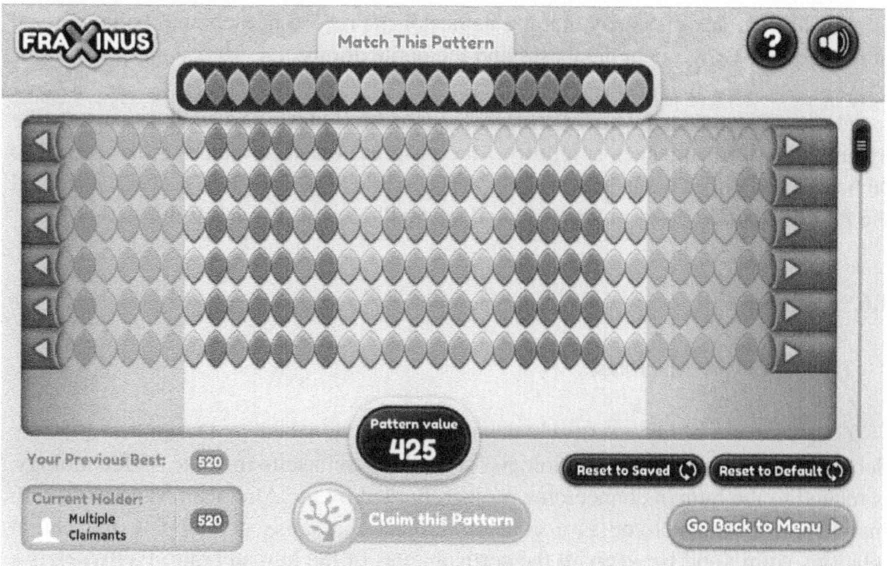

Figure 1.6. *Fraxinus: sequences almost aligned. Compared to the previous figure, it can be seen that the blocks of colors of the six sequences are almost perfectly aligned, both among themselves and with the model. But although the score has gone up from* -148 *to* $+425$*, the alignment can still be improved since it can be read, in the bottom left, that the record to beat is 520*

Interactivity – The game, accessible only via Facebook, benefits naturally from the facilities of the social network to attempt to develop interactivity between players and especially encourage them to popularize the game among their *friends*. The player is strongly encouraged to invite *friends* to play, to compare his/her score with those of his/her *friends*, to share results and of course to talk about the game on his/her *wall*.

Challenge/emulation – When the player initiates an alignment *puzzle*, the *record* to beat appears in the game window with the name of its holder. It suffices to just match it to seize it, and then the player is asked to defy his/her *friends* to obtain a better score with the same *puzzle*. The player can either start a new *puzzle* or try to "steal" a *record* held by someone else by improving the alignment. All the completed *puzzles* appear in a pane in the shape of icons, mentioning the score obtained and displaying the words *improve* or *try again* depending on whether the player has beaten the *record* or not. It is exciting to hold a *record* and obviously the game tries to enhance this aspect and arouse the desire to prove either that the player can do better or that he/she is the best.

In addition, designers say that the names of the record holders will be mentioned in the sequence alignment databases and the publications.

Design/feeling – The interface is agreeable, clear and well arranged; the manipulation of sequences is intuitive; the progress bar is stimulating; the record to be beaten is highly visible and the help is accessible. The design is well achieved and the four nucleotide bases are represented by small leaves of four harmonious colors.

1.6. Eyewire

Type of task – Macrotask.

Goal/launch – EYEWIRE[11], developed and launched in December 2012 by Sebastian Seung's team, neurobiologist at the Massachusetts Institute of Technology, is related to the human connectome project, funded by the American National Health Institute. The connectome is a colossal project that aims at mapping the neural relations within and between all the cortical areas of the human brain. EYEWIRE is a game designed to help the reconstruction of the extensions of neurons in 3D from photographs of serial histological sections of the retina. From the 3D mapping of neurons, neuroscience researchers will be able to identify the synapses and relate the connections and brain activity. Seung and his team have just published in *Nature*, with *The EyeWirers* as co-authors, important results on the visual perception of motion through the data provided by the game [KIM 14]. This advance validates, according to them, the use of crowdsourcing through the online game to advance major scientific issues quickly and efficiently.

Audience – In May 2014: More than 135,000 recruited players and 150 countries represented (source: Wikipedia[12]).

Benefits of crowdsourcing for the problem to be solved – The eye and human intelligence combined together prove to be much more efficient than computer programs in detecting boundaries and cell contours. This is particularly true when the extensions of nerve fibers are entangled in a dense network, as is the case in the retina. The player is faced with a *cube* that represents a piece of retina traversed by a portion of neuronal branch that has been partially and roughly automatically delimited by image analysis. The player must improve the tracing by *coloring*, which he/she identifies as belonging to this neuron on a electronic microscopy photograph that represents the same *cube* in cross-section. Each *cube* is decomposed into 256 slices. The arrow keys of the keyboard allow the player to move through the

11 http://eyewire.org/.

12 http://en.wikipedia.org/wiki/EyeWire.

thickness of the *cube* jumping from one slice to another and *to follow* the neuronal branch by correcting the work of the analysis algorithm. This is achieved by *coloring* the portions of neurons possibly omitted, or even by detecting ramifications also ignored by the computer. A whole neuron corresponds approximately to 1,000 *cubes*, each of these 1,000 *cubes* is analyzed by 3–5 different players, and it is the correlation between their tracings that will legitimize the final mapping.

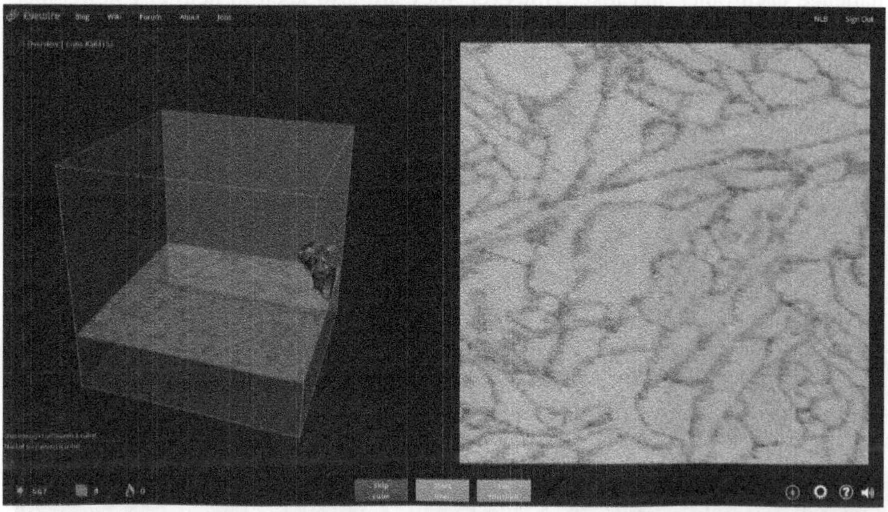

Figure 1.7. *The interface of Eyewire. On the left, the "cube" (fragment of a histological section of the retina) which contains the neural branch to be outlined, of which a fragment can be seen. On the right, the corresponding transversal cross-section where the player "colors" areas that seem to him/her to belong to the neuron. He/she, therefore, "follows" the neural branch by traversing the entire thickness of the cube, through which he/she moves with the arrow keys of the keyboard*

Required human skills – Designers insist on the fact that to become an efficient *Eyewirer*, no knowledge in neurology or even in general biology is required. The game is available to anyone with a good sense of observation and spatial visualization.

Player's education – Even if no knowledge in neurology is required, the players have access to a wealth of scientific information about the project in which they participate by playing. They are also regularly kept informed not only about the evolutions of the game, but also about the resulting scientific progress. On the other hand, elder and experienced players are given the power to supervise the work of the community. Thus, they are recognized as having a true expertise in the visual ability to detect the contour of a neuron in an electronic microscopy picture, that is a document which most ordinary people only see as a patchwork of gray shades.

Interactivity – Although the task is rather monotonous and each player colors his/her little piece of neuron on his/her own, conviviality and interactivity seem highly developed in the game. An instant messaging window is available in the interface, and the players extensively debate and seem to form a large community of regulars that challenge each other and joke. However, they are also keen to help any newcomer who asks a question. On the other hand, the messaging window also accounts for the activity of the game and gives an idea of the number of players connected by communicating in real time the points obtained by them. This is one way to stimulate emulation and an opportunity for them to congratulate each other.

Figure 1.8. *Eyewire. This image from the blog of the game highlights a synapse (the point of junction between two neurons) discovered by means of the activity of the players*

Challenge/emulation – Given the relative aridity of the task (once again, the objective is about scanning grayish pictures looking for continuity and homogeneity), one feels the wish to make it more interesting and encourage players to compete by rewarding them through a rather complex points, titles and classification system. The player receives all the more points when his/her tracing is close to the average of the tracings of the other players. He/she is also rewarded by a *pro rata bonus* of the time spent, intended to reward its application and the care taken in his/her *coloring*. For an equal amount of work, some cells are rewarded twice as much because they are deemed more difficult to isolate (*starburst cells*). In addition to the points, the meticulousness of the player and the relevance of his/her tracings are evaluated by an accuracy percentage. Once the player has reached an accuracy of 80%, he/she is allowed to become a *pioneer*, that is to be the first to intervene on a *cube*, and thus enjoys a *clearing bonus*; he/she then wins points again whenever another player deals with this *cube*. The players who have demonstrated their skills and pass the tests can be promoted to the ranks of *Scout* or *Scythe*, which gives them the rights to supervise the work of other players and to validate or invalidate some tracings. Thus, the complexity of the classification, the diversity of the evaluated parameters, the seniority as well as the difficulty of the processed cells create a hierarchy within the players and maintain the emulation. In addition, the regularly organized competitions are a way to retain players, who have the opportunity to compete over a specific goal. These *challenges* are of various kinds: for example, *Happy Hours* are held weekly and consist of earning the highest possible number of points during this period of time, with bonuses as reward for the winner. *Marathon* is a challenge: it is necessary to complete an entire cell (i.e. 1,000 *cubes*) in a given time (currently 26 h). *Trivia* is a type of multiple choice questions (MCQ) quiz; the *Diabolic Cube* challenge consists of coloring in the best possible manner regions that are particularly difficult. This difficulty is due to the fact that the neuronal branches of different cells are closely entangled. These one-off events are elements that spice up the game by introducing competition and suspense. Since May 2014, players can form teams, which gives rise to an inter-teams classification. Finally, *Lore of Kor* was launched recently (3 June 2014). It is a neuron-coloring contest based on the background of a science fiction scenario, in which two teams confront each other, with premiums and various bonuses up for grabs. The developers keep breathing life into the game; to this end, they take great care to minimize the monotony of the task. This is achieved with the help of all kinds of devices meant to promote the interactivity and competitiveness, while creating, maintaining and strengthening the sense of belonging to a community.

The other method of securing loyalty and motivation involves an extensive use of the very promising concept of *citizen science*, which results in an extreme valuation of the player as the driving force of the research. Researchers and developers frequently communicate on the blog of the game and do not miss an opportunity to associate the players, known as *The Eyewirers*, with the progress and the results obtained [KIM 14]. The other method of fostering loyalty and encouragement is to *cajole the player* and strengthen, through the granting of titles and diplomas with esoteric soundings, and his/her sense of belonging to the very closed brotherhood of the trailblazers of neural networks.

Design/feeling – The game interface is neat, clear and pleasant. The black background highlights the modeling of neurons in 3D which, with all their branching extensions, give rise to quite spectacular figures that can easily be rotated in all directions. On the home page, a table summarizes the different modes of classification (daily, weekly, monthly, individual and by team). In the actual game interface, the view of the *cube* in 3D is manipulable and adjustable, and it is quite pleasant to gradually see the volume of the neuronal branch that the player is coloring over the two-dimensional (2D) cross-section appear in 3D.

1.7. Citizen sort

CITIZEN SORT[13] is a portal for *citizen science* designed by a group of researchers and students from the School of Information Sciences of the University of Syracuse, in the state of New York. The proposed games have been developed in collaboration with biologists and naturalists. The objective of this initiative is twofold:

– to generate the data relevant to biologists/naturalists (especially in entomology, botany and oceanography) through the games;

– to help researchers in information science and specialists in the field of human–machine interactions to understand how gaming, science and citizenship concepts interact, and to assess the role of gamification as an engine for motivation. The portal CITIZEN SORT allows studying how different types of games, addressing different types of players, can generate different qualities of data.

The portal CITIZEN SORT offers games to classify various species of plants, insects and animals. Two games are currently available: HAPPY MATCH[14] and FORGOTTEN ISLAND[15]; a third game, MARK WITH FRIENDS[16], is in development. The portal CITIZEN SORT claims through the general classification more than 3,000 players, but only 1,200 are active (with a positive score).

1.7.1. *Happy match*

Type of task – Microtask.

Goal – This consists of helping the classification of living beings by characterizing photographed specimens according to different criteria. In practice, a series of 5–10 pictures is proposed, and the player will have, for each picture, to give his/her opinion on several morphological criteria by assimilating the specimen to a form of reference, of which a characteristic example is presented below. For

13 http://www.citizensort.org/.

14 http://www.citizensort.org/web.php/happymatch.

15 http://www.citizensort.org/web.php/forgottenisland.

16 http://www.citizensort.org/web.php/markwithfriends.

example, if it is about butterflies, the goal will be to assimilate each picture to a general type depending on the shape of the wings at rest, then on the dominant color of the wings and then on the shape of the pattern of the wings, etc.

Thus, the morphological data collected by crowdsourcing contribute to formally identify the sample at the specific or even at the subspecific level. Since the pictures are dated and show information about the location (by geolocalization), the correlation of the taxonomic information with the date and the location can, for example, inform about the impact of certain forms of urbanization on an ecosystem, or highlight climate change, a change in biodiversity, the appearance of an invasive species, etc. Currently, it is possible to play with pictures of moths (HAPPY MOTHS), sharks (HAPPY SHARKS) or rays (HAPPY RAYS).

Benefits of crowdsourcing – Crowdsourcing is used here to achieve a very heavy task in terms of time and data volume to process. As often in the case of microtasks, it is the redundancy and the aggregation of the data provided by the players who, on the condition that they are sufficiently numerous, will provide quality data as good as the annotations of experts.

Required human skills – These are minimal and amount to a sense of observation, an interest in nature, patience, etc.

Player's education – By repeatedly playing, and by using the provided correction elements (see challenge/gamification), the eye can be exercised and one can learn to properly discriminate between some morphological elements. These elements are at the base of the determination keys, and the player can become unbeatable about the family, genus and species Latin names. It can reasonably be assumed that this kind of game delights people who already have expertise in taxonomy (e.g. amateur entomologists).

Interactivity – This is not developed in HAPPY MATCH. The player characterizes his/her images in his/her corner, without contact with what the other players do, without a discussion space such as instant messaging or a forum. On the other hand, users can read on the portal CITIZEN SORT a notification about an upcoming game, MARK WITH FRIENDS which seems, as its name indicates, to place interactivity at the center of the data acquisition strategy.

Challenge/gamification/interface – The interface is visually appealing and one can feel that it was designed with the help of designers. Compared to other games of image annotation (e.g. those of the ZOONIVERSE portal), it can be noted that there is a very pronounced effort to develop the entertaining and esthetic aspects. This starts from the interface which is organized to present the task to achieve in a harmonious and ergonomic way. Skillfully, the essential element of gamification is also the parameter that makes it possible to assess the quality of the submissions and thus to validate the data provided by the player: in the series of the proposed images, one or two pictures of individuals already known and identified with certainty are *hidden*. If the player correctly classifies these two images on all the proposed morphological parameters, it could reasonably be estimated that the classification that he/she performs on the other images is reliable. At the end of the game, the result reveals if the two known images have been correctly classified on all the parameters; in this case, the player wins *bonus* points and the two species appear as specimens collected in his/her profile; otherwise, for each wrongly assigned morphological characteristic, the correct decision appears, with regard to the picture of the specimen. This allows him/her to know where and why he/she was wrong, and therefore adds an educational element to the ludic aspect.

Figure 1.9. *Citizen Sort: Happy match. The first screen of the game presents 10 pictures of butterflies. At this stage, the player must just validate the pictures, that is to eliminate those that are blurry, truncated or of too poor quality to be used. In the following screens, he/she will categorize butterflies following various morphological criteria (wing shape, dominant color, etc.)*

1.7.2. *Forgotten Island*

Type of task – Microtask.

FORGOTTEN ISLAND is the *adventure game* version of HAPPY MATCH. The requested task is the same (classify images of animals according to morphological criteria), but in a context that mimics adventure games: one plays the role of an adventurer lost on an island where an explosion has ravaged a laboratory and scattered specimens of animals, which must be identified and reclassified. The character that the player embodies moves in a scenery and can glean objects needed for his/her mission, but can also teleport himself/herself from one strategic location of the island to another under the orders of a robot that tells him/her what to do and where to go.

1.8. The Nightjar project

Goal/background – The NIGHTJAR PROJECT[17] is a research program launched in August 2013 by the Sensory Ecology Group of the University of Exeter, in collaboration with the Behavioural Ecology Group at the University of Cambridge. The theme concerns the study of the strategies of camouflage in nature, relative to survival, and the model is that of the *nightjar*, a bird which breeds and lays eggs on the ground. Photographs of nightjars in their natural environment are modified according to the visual parameters of their natural predators (different from ours) to compare the appearance of eggs, chicks and the adults to their natural environment in order to quantify the effectiveness of the camouflage. The study focuses on various species of nightjars, all highly depending on appropriate camouflage strategies to escape their predators. Other birds are also studied (plovers and coursers), but nightjars constitute a model of predilection because eggs, adults and chicks demonstrate different strategies of camouflage. The team of researchers, in collaboration with a game designer, has developed three games to test the effectiveness of the strategies of camouflage of the eggs and of adult nightjars, depending on the type of visual perception of their various predators.

Benefits of crowdsourcing for the problem to be solved – The problem to be solved is the evaluation of the effectiveness of the camouflage strategy, which can be measured by the average time to locate a nest or a bird in a setting. The player will choose to embody a predator category and seek to locate the nest or the bird by watching photographs taken in natural sites. These photographs have been modified to correspond to the di-, tri- or tetrachromatic vision characteristics of each of the three types of natural predators of the nightjar. The data retrieved by crowdsourcing are, therefore, location times, which are closely correlated with the effectiveness of the camouflage.

Interactivity/player's education – The proposed type of game does not lend itself to interact with other players. However, a praiseworthy effort from the designers can be noted to explain the nature of their work: accurate and detailed scientific information

17 http://nightjar.exeter.ac.uk/.

on the project is easily accessible via the website of the project, while more technical data concerning the games are provided on the blog of the designer of the game[18].

Figure 1.10. *The Nightjar project. The introduction screen of the Nightjar game, one of the games of the project*

1.8.1. *Nightjar game/Nest game*

Type of task – Neither microtask, nor macrotask.

Goal – WHERE IS THAT NIGHTJAR? (NIGHTJAR GAME)[19] is the first game launched in August 2013, followed in November by WHERE IS THAT NEST? (NEST GAME)[20], which works on the same principle. The player must, in a limited time, locate either a bird or a nest in the picture that is presented to him/her after choosing

18 http://www.pawfal.org/dave/blog/.

19 http://nightjar.exeter.ac.uk/story/nightjargame.

20 http://nightjar.exeter.ac.uk/story/nestgame.

the type of predator (monkey = trichromatic vision or mongoose = dichromatic vision) and its type of vision. The timing that it takes to locate (or not) the nest or the bird is directly correlated with the effectiveness of the camouflage strategy.

Challenge/gamification – After accepting that the data of the game are employed in the context of scientific work, after indicating his/her age group and after choosing what predator the player was embodying (which conditions the colorimetrical parameters of the presented pictures), part of the game consists of looking for a bird or eggs in 20 successive pictures. The main ludic element is the task itself: it is fun to find a bird (or eggs) camouflaged in a natural scenery and not necessarily easy to locate it when the landscape is seen in shades of green, that is with the dichromatic vision of a mongoose. The limited time (30 s) spices up the task. If the player clicks on a location where there is nothing, a large red cross appears accompanied by a low-pitched sound. If nothing was found at the end of the allotted time, the solution appears in the form of a rectangle around the camouflaged element, and the failure is hailed with a humorous mention "You will be hungry, tonight!".

Figure 1.11. *Nightjar game. A nightjar (designated by the arrow) is hidden in this setting, and the player has 30 s to locate it and click on it*

1.8.2. *Egglab game*

Type of task – Neither microtask, nor macrotask.

Audience/popularity – The blog of the designer of the games[16] reported, in May 2014, that 9,000 players had played 20,000 games and tested 400,000 shell patterns representing more than 30 generations of artificial eggs. The average overall time of locating an egg has increased by approximately 0.4 s since the beginning of the experience, initiated in April 2014; the launch date was appropriately chosen in connection with Easter.

Principle and goal – EGGLAB is a piece of software designed to simulate the genetic variability that characterizes the visible patterns on the eggshells, which play a crucial role in camouflage and are, therefore, directly correlated with predation. The simulated eggs are images that are juxtaposed with pictures of different laying sites, and tested through the game EGGLAB GAME[21]: the player has 30 s to locate the egg in the scenery, and the less noticeable patterns are reinjected in the program to generate a new series of sets. This results in shells increasingly more resembling the ambient scenery, in a similar manner to what happens in nature with the process of natural selection.

Challenge/gamification – The game is built on the same principle as the previous two, with the difference that this time synthesis eggs (therefore drawings) must be identified in a natural setting. The natural setting is a picture of the habitat of one of the three species of the studied nightjars. The presentation and the scenery look more like a *cartoon*. The player is asked to enter his/her name by clicking successively on eggs carrying letters. At the end of the game, the average time flashes, and the player is warned that due to him/her, the eggs will become increasingly difficult to detect. He/she is invited to continue to play to see how the difficulty will evolve, which is a way to try to catch his/her loyalty. A clever way to attract a larger number of players is the recent possibility for the player to hide himself/herself the eggs in the setting (*Making an Easter egg hunt*) and to send his/her *Easter eggs hunt* to his/her friends. To this end, he/she is invited to choose a natural setting (picture), and then to slide or place virtual eggs with various shell patterns, selecting himself/herself the place of the scenery where the camouflage seems optimal. When it is finished, a click on *send* generates a url that can be copied and sent by e-mail.

---

21 http://nightjar.exeter.ac.uk/egglab/.

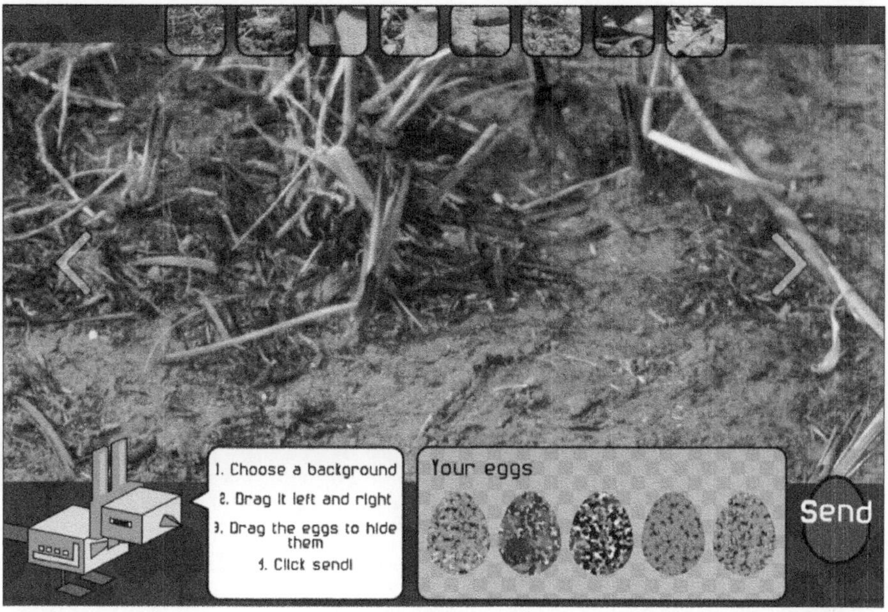

Figure 1.12. *The game Egglab.* Making an Ester egg hunt *is a variant of the* EGGLAB *game: the player chooses a setting, and then hides himself/herself the eggs by clicking and dropping them in areas of his/her choice. He/she can then send the url of the created game to his/her friends*

1.9. References

http://www.pri.org/stories/2014-05-12/your-video-games-may-revolutionize-health-and-scientific-research.

http://boinc.bakerlab.org/.

http://singularityhub.com/2011/09/22/in-3-weeks-video-gamers-defeat-biochemical-puzzle-that-scientists-couldnt-solve-for-years/.

http://www.pri.org/stories/2014-05-12/your-video-games-may-revolutionize-health-and-scientific-research.

http://biomedicalcomputationreview.org/content/biology-game-crowd.

http://www.pourquoidocteur.fr/Jouez-pour-aider-la-recherche-sur-les-maladies-genetiques----4090.html.

2

Games with a Medical Purpose

When it comes to resort to crowdsourcing through gaming, the medical sector is also very active, whether at the research and the development of innovative processing techniques level, or that of support to diagnosis. We have deliberately ruled out one of the crowdsourcing activities for medical purposes which is among the most popular, CELL SLIDER [GOO 13], proposed by the ZOONIVERSE portal, because it is devoid of any ludic aspect, even vaguely. It should be noted that among the games described below, whether it concerns microtasks or macrotasks, the skills required from the players are very uneven. They range from the simple ability of observation (WORM WATCH LAB, MALARIA TRAINING GAME), to a scientific, genetic or highly developed medical culture that can be very extensive (DIZEEZ, THE CURE). The skills required may be dexterity and the typical speed of console gaming adepts (PLAY TO CURE).

2.1. Nanodoc

Type of task – Rather macrotask, but issues remain in the use.

Goal/launching/authors – NANODOC[1] is a game that was launched in September 2013 by Sangeeta Bhatia, researcher at the Koch Institute for Integrative Cancer Research at MIT, and his team. It concerns nanomedicine and is aimed at the design and development of nanoparticle-based treatments. Nanoparticles are currently studied in a large number of research laboratories around the world because they represent a considerable hope for cancer treatment and more generally for all diseases that are resistant to conventional therapies. One of the major interests of nanoparticles is their ability to perform the treatment on the site of the tumor itself: an ideal size can be given to them such that after injection they circulate in the blood

1 http://nanodoc.org/.

vessels and can only escape at the place where some looser vessels irrigate the cancerous tumors. As a result, the nanoparticles carrying drugs will accumulate passively at the location of the tumors while avoiding damage to healthy tissue.

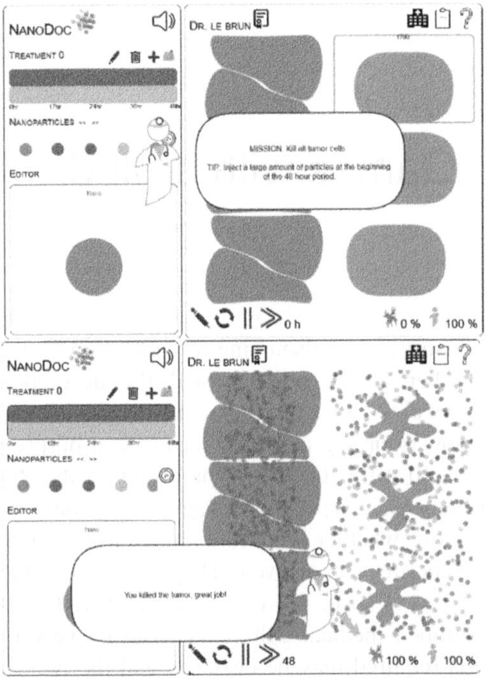

Figure 2.1. *Nanodoc. The objective is to build the combination that will fulfill the mission shown in the left figure by using the nanoparticle editor. On the right, the result (success): all the cancer cells (in gray) have been destroyed. For a color version of this figure, see www.iste.co.uk/lafourcade/games.zip*

Audience – A source accounts for 50,000 simulations carried out by 2,600 players in October 2013, or 1 month after launching. It is impossible, in the interface of the game, to find the number of registered players.

Benefits of crowdsourcing for the problem to be solved – There are multiple ways to design nanoparticles and multiple parameters that influence their behavior. The challenge is to understand in the design of the particle what will allow the optimization of the issuance and the efficiency of the treatment. This is achieved by changing its mode of action and its interactions with other particles and with the physical environment of the tumor area. The aim is also to ensure that nanoparticles act as an organized system (such as exists in nature in social insects, for example) in order to synergistically enhance their therapeutic effect.

Thus, there are thousands of combinations of action modes, of possible therapeutic strategies, and it would take a considerable amount of time to review and carry out the tests and trials necessary to validate them. Moreover, each problem is different, which excludes programming a computer that would automatically design nanoparticles. That is why Sangeeta Bhatia's team has designed a simulator that allows the modeling of how nanoparticles interact with each other and with the tumor. The idea is to turn this simulator into a gaming interface: on the one hand, bioengineers from the research team design various scenarios of tumors, while on the other hand, players will imagine various treatment strategies by designing virtual nanoparticles that will be tested with the simulator.

The idea is that the most promising strategies are validated by the research team to serve as a basis for *in vitro* testing.

Required human skills – Initiative, curiosity, inventiveness, tenacity, patience, etc.

Player's education – From the beginning of the game, the player is instructed through a tutorial about the various parameters inherent to the creation and use of nanoparticles (useful concentration, specificity, particle size, targeting, etc.) and brought to test various assemblies and configurations in order to kill all the tumor cells, leaving healthy cells intact. Each parameter is explained briefly, and then the player is invited to put it into practice by developing a treatment capable of completing the proposed objective. The problem comes from the fact that the explanations are too brief and the player finds himself/herself executing dozens of attempts without necessarily finding the right combination. Because the player at this point is still in the tutorial, he/she might expect that in the event of repeated failures, the solution should be given instead of the formula "you did not kill any cancer cells. Try to improve the treatment". In case of treatment failure, it is still possible to go to the next level, but it is extremely frustrating, and the problem is a recurring one. The *learn* section of the game brings a lot of rather well popularized information about the real biochemical concepts involved in the different treatment strategies illustrated by the tutorial. However, it does not absolutely help the player to understand why what he/she tries does not work. Unmotivating…

Interactivity – Obviously, NANODOC is a solitary game. Competition with other players was not foreseen, and the opportunities for interaction are limited to a blog where, apart from the presentation of the game, no comment is available. The *learn* category is also presented as a blog, and the comments can be counted on one hand.

Challenge/emulation/gamification – The gamification essentially consists of calling the player *Doctor*, to congratulate him/her when he/she has succeeded and to award him/her with a *certification* indicating the skills acquired during the whole tutorial. No emulation since the gamer plays alone, no satisfaction other than reaching the target and destroying the tumor cells while preserving the healthy cells. However, the design of the game is rather attractive; the interface is sober, the animation illustrating the impact of the treatment is pleasing although a bit long, but this is certainly deliberate and designed to create a bit of suspense: "will we succeed in eliminating all the tumor cells?"

Conclusion – The game is promising but disappointing, the tutorial is badly made, it fails completely in its purpose since the player is blocked during the learning phase and discouraged before getting to the heart of the matter. There is no ranking, no interaction with other players and no emulation... This is far from the feeling of EYEWIRE, while paradoxically the task at hand seems infinitely more interesting and varied.

2.2. Dizeez

Type of task – Microtask, but reserved for specialists.

Background/presentation – DIZEEZ[2], launched online in October 2011 by the Su Lab, a California laboratory specialized in the adaptation of computer tools to the problems of biological research, challenges those who think that they have knowledge on the genetic basis of human diseases. The principle is simple: the name of a gene is given (for example: IYD (iodotyrosine deiodinase)), and it must be associated with one of the five proposed diseases within a time limit. Each correct answer gives points.

Problem to be solved and benefits of crowdsourcing – In general, the associations between genes and diseases are correct in the databases. By analyzing the data from the game, it is therefore expected that players' responses confirm what is already established.

But if enough people play, it is also presumed that gene–disease associations that are missing in databases will therefore be revealed. Thus, the game has the vocation to narrow the gap between the knowledge available in the biomedical literature and that of databases, usually incomplete, by filling the gaps in the databases of annotation of genes. According to the designers, one month after its launch, DIZEEZ has already led to the identification of new gene–disease associations which are well defined in the literature. However, they may not be taken into account in biostatistics computer analyses since they do not yet appear in databases.

The designers are aware of the need to bring improvements to the game, but are very excited by the potential of this type of approach (GWAP) to help in structuring

2 http://sulab.scripps.edu/dizeez/.

the knowledge. Nonetheless, they have not published any new posts in the associated blog since March 2012.

Required human skills – Fine-grained knowledge of the genome and gene/disease associations.

Player's education – The player does not learn anything, on the contrary, he/she must be properly educated in order to play efficiently. Not knowing the genes by their scientific name is a limitation and reduces the experience to a simple game of chance.

Gamification/interface/critical – Many of the ingredients that would make a game addictive are present (limited time, gain/loss of points system, ranking, top 10), the interface shows an effective sobriety, no frills, few colors, and we feel that *we are not here to have fun*. However, this game is for doctors, geneticists, actors in the research on genetic diseases, biochemists and certainly not for the general public. Who in the general public, even with knowledge of biology or genetics, can possibly know that the HOXB7 gene is associated with leukemia, or the BCL10 gene with lymphoma? DIZEEZ is a game made by specialists for specialists. It is an elitist version of crowdsourcing that actually concerns only a limited number of players.

2.3. The Cure

Type of task – Unclassifiable, with a rather macrotask tendency, intended for specialists.

Goal – The objective of the game is to build an effective model for the prediction of the survival time for breast cancer based on the gene expression and on the information from tumor samples. THE CURE[3], developed and launched in September 2012 by the same laboratory as DIZEEZ, is presented as a poker game in which the cards are represented by genes: the point is to assemble the best *hand* by selecting the genes that compose it to beat *Barney*, an automaton that offers a combination calculated from data relative to known survival rates. Each added *card* (gene) varies the score and, ultimately, the assembly of 5 genes gives a higher or a lower score than that of *Barney*. The goal is to use genetic data relative to real clinical cases and to correlate them with what is known of the evolution of the disease in these patients. This will derive predictive models and information about the involvement of such a gene or of such a combination of genes in the capacity of a tumor to metastasize, for example.

3 http://genegames.org/cure/.

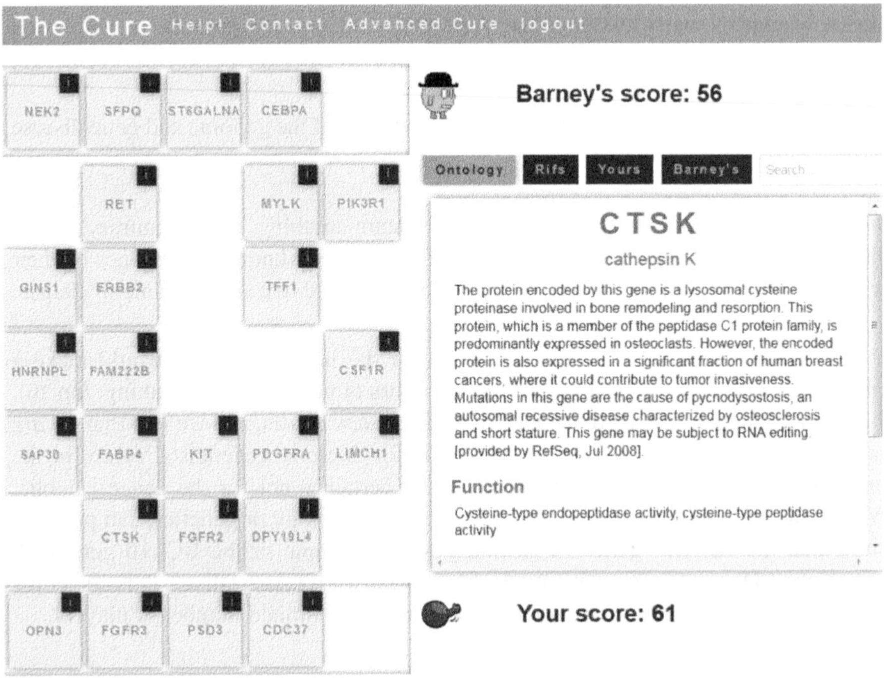

Figure 2.2. *The Cure. By analogy with poker, it is necessary to constitute a* hand *by choosing 5 genes in the central part. The selected genes appear in the top window, and the text on the right gives information about the gene when the mouse passes over it, in order to guide the choice. Each gene chosen contributes to the scores, the goal is to beat that of Barney (the algorithm of choice)*

Audience – Between September 2012 and September 2013, 1,000 players have registered and played almost 10,000 games, according to [GOO 14].

Benefits of crowdsourcing for the problem to be solved – Searching for predictive models is faced with two problems: the uncertain reproducibility, due to natural biological variability, but especially the impossibility to test all the possible combinations of genes because of their too significant number: a unique combination of 25 genes may be an excellent predictive model, but with more than 20,000 genes an order of 10^{82} combinations of 25 genes can be obtained. Hence, the interest of resorting to intelligence and to human knowledge to develop combinations associating genes not randomly as a machine would, but guided by knowledge based on genes, by biological mechanisms involved in cancer, and by clinical data on the disease. Thus, the player is invited to compose his/her *hands* according to his/her knowledge if he/she has any, or by searching for the role of any one gene on the Web.

To somewhat facilitate the selection, information on each gene, whenever it is known, is available in the game interface. The combinations developed by players will feed learning algorithms that will enable the development of predictive models for the prognosis of breast cancer.

Required human skills – Playing requires some knowledge in genetics/medicine/biochemistry, which seems to be hardly compensable by the curiosity needed to find on the Web information relative to genes known for their involvement in certain cancers, so as to be well informed in order to constitute the *hand*. In the FAQ, it is nevertheless stated that the higher the knowledge in genetics and in cancer, the easier it will be to play. However, it is also mentioned that it can be acquired during the game, since a lot of scientific information is provided in the game, as well as links to Web resources.

Player's education – The player learns by playing if he/she has the patience to read and to assimilate all the information relating to the genes that is proposed to him/her in the interface of choice. It seems a bit contradictory to the principle of a poker game, which assumes playing the matches in a quick succession and trusting (at least partially) in chance. Moreover, the *random* choice of genes to constitute the *hand* gives rather good results because in doing so, *Barney* can very often be beaten. However, the designers claim that by testing the predictive models from game data from sets of real data, it is easy to distinguish between serious players and those who have randomly assembled genes. Futhermore, the predictive models originating from the game would clearly demonstrate the acquisition of skills among players, and give results comparable to those obtained by conventional methods [GOO 14].

Gamification/Design – The game interface is sober and appealing, and the ergonomics is pleasant. Each game is fast, and the evolution of the two scores as the adjunction of genes is carried out maintains the suspense. The accumulated scores give rise to a ranking between players.

2.4. Malaria Training Game

Type of task – Microtask.

Goal – Launched on April 25, 2014 (World Malaria Day) by Biogames, Ozcan research laboratory department at UCLA specialized in the production of games for medical diagnosis. MALARIA TRAINING GAME aims to recognize by crowdsourcing cells infected by malaria. This is in fact a training module in which the player is invited to examine pictures of isolated single cells and to characterize each of them as *healthy*, *infested* or *questionable*. Each game consists of diagnosing about 500 cells, which have already been characterized by professionals. A score is awarded at the end according to the number of correctly designated cells, but also to the number of *false positives*, *false negatives* and cells wrongly characterized as *questionable*. The educational dimension of the game is to show to the player for what cells and in what sense (*false negative*, false positive, *false questionable*) he/she was wrong, so that he/she can improve in the next game.

The long-term goal is twofold: (1) that this ludic training module be widely used to improve the training of medical personnel and the quality of the diagnosis in all the areas where malaria is rampant and (2) that it be useful for educating learning algorithms with the objective to automate diagnosis.

The interest of this kind of approach is such that it is adaptable to other diseases, and that it can also be used to measure parasitemia in patients under treatment.

Benefits of crowdsourcing for the diagnosis of malaria – A pathologist must inspect at least 1,000 red blood cells (RBCs) in order to establish a reliable diagnosis for the malaria. This is a long and tedious task, where error is frequent: according to the designers, 60% of reported cases are *false positives*. Any reduction in the number of *false positives* can avoid unnecessarily costly treatments and hospitalizations. In their published work, the designers believe to have shown that by employing the visual qualities and the learning abilities of the public through games that combine learning and feedback, they could obtain a reliable diagnosis from digital medical samples. In the case of binary diagnosis (infested/healthy), the accuracy and reliability of the experts can be approached, since the diagnoses established through the game differ by 1.25% from those of qualified professionals [MAV 12].

Required human skills/education – Sense of observation, desire to improve by using the correction to detect and eliminate sources of errors. It is obvious that by practicing the game, and starting from absolute ignorance, we can quickly improve. The game requires no prior knowledge, in particular, no skill other than that to be able to recognize infected erythrocytes.

Challenge/gamification/ergonomics – The game does not require registration. A game (500 cells) takes a few minutes, it is fast enough to raise the desire to try to improve the score one more time. Cells are presented on a grid of 5 by 7, or 35 cells to be characterized per screen. The characterization of cells is intuitive, it is done by a simple click after selecting the suitable tool. Two progress bars indicates: one bar indicates the score, in green if correct answers predominate, in red otherwise, the other bar indicates the percentage of examined cells. The interface is simple, clear, unadorned; two buttons allow the adjustment of the background sound of the game. At the end of the game, the player can acknowledge his/her mistakes, and submit his/her score in order to appear in the rankings.

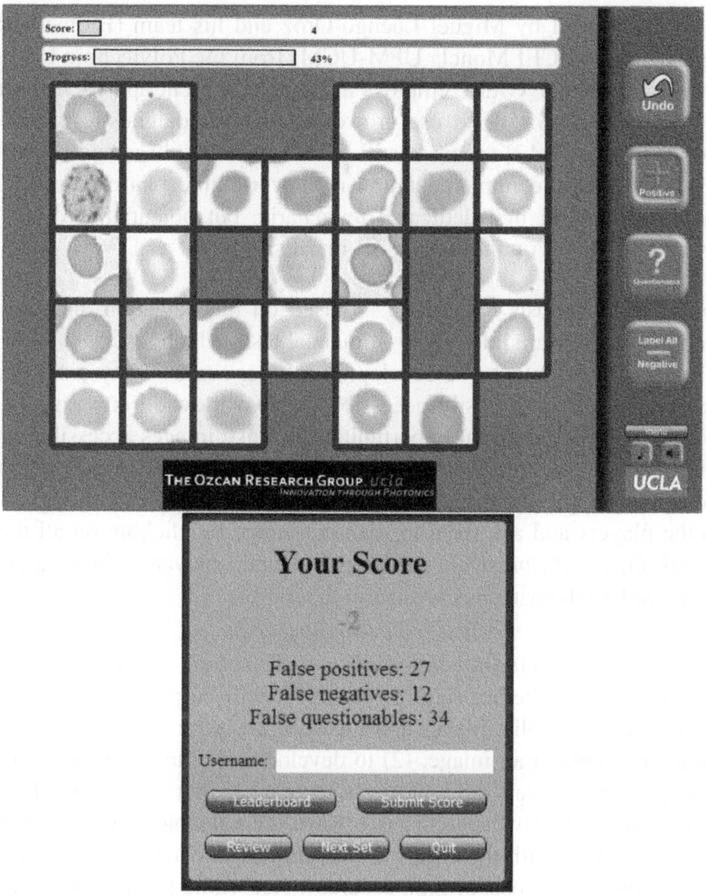

Figure 2.3. *Malaria training game. Top figure: the cells are characterized (*positive, negative *and* questionable*) by clicking them after choosing the appropriate tool in the menu on the right. The figure on the bottom shows the score at the end of the game*

2.5. Malaria Spot Game

Type of task – Microtask with the slogan: "a minute game against malaria".

Audience – One month after the launch of the game (April 2012), players from 95 countries have completed more than 12,000 games and generated a database of more than 270,000 clicks on the test images. According to [LUE 12], in April 2013, more than 15,000 people from 100 different countries have generated almost 1,000,000 clicks.

Goal – Developed by Miguel Luengo-Oroz and his team (Biomedical Image Technologies Group & CEI Moncla UPM-UCM) from the Polytechnic University of Madrid, MALARIA SPOT GAME employs a slightly different approach from that of MALARIA TRAINING GAME to use crowdsourcing for diagnostic purposes. The game described in [LUE 12] seems to still be in its testing phase and constitutes the preliminary study for the implementation of a worldwide platform specialized in online diagnosis. It is at first dedicated to malaria, but potentially to any disease whose screening is based on the analysis of images.

Currently, the diagnosis of malaria is based on the detection and counting of parasites on a blood smear slide. This is a manual task, which takes about 30 min for a specialist. Hence, the interest of developing a faster screening system, as efficient as a specialist's work, and easily accessible. The worldwide coverage of mobile telephony and the Web represent opportunities to make images accessible anywhere in the world. Therefore, after having seen a few, it is fairly easy to identify parasites on a blood smear image. The principle of the game is thus to present images of blood smears to the players and ask them to scan the image by clicking on all the shapes that they assimilate with *plasmodium*. In this manner, players achieve a count, the goal being to find all the parasites present in the sample.

Benefits of crowdsourcing for the problem to be solved – From the analysis of all the players' *clicks*, whether right or wrong, it will be possible: (1) to evaluate the speed of a non-specialist player and the accuracy with which he/she is able to detect pathogen agents in an image, (2) to develop a way to combine the results of the different players to obtain a diagnosis as accurate as that of a qualified expert. As a result, the benefit of crowdsourcing is to save the valuable time of specialists by replacing each of them with thousands of players whose counts will be combined to result in a collective diagnosis of high accuracy. According to Miguel Luengo-Oroz, 3% of the time that the world's population spend playing video games would be more than sufficient to diagnose all the cases of malaria in the world . The designer also mentions the possibility of using the data generated by the game to automate diagnosis and to minimize the time required for counting. This data could also be employed to develop a system of microscopy image acquisition by mobile phone to allow the rapid transfer of images to the platform of remote diagnosis. The results have showed that by combining 22 non-expert games ensures accuracy in the enumeration of parasites in the order of 99%. The same result is obtained by combining 13 games of trained players.

For the purpose of testing and validation of the method, the images proposed to players have already been characterized by professionals. However, it is intended to subsequently add images not yet diagnosed to increase the database.

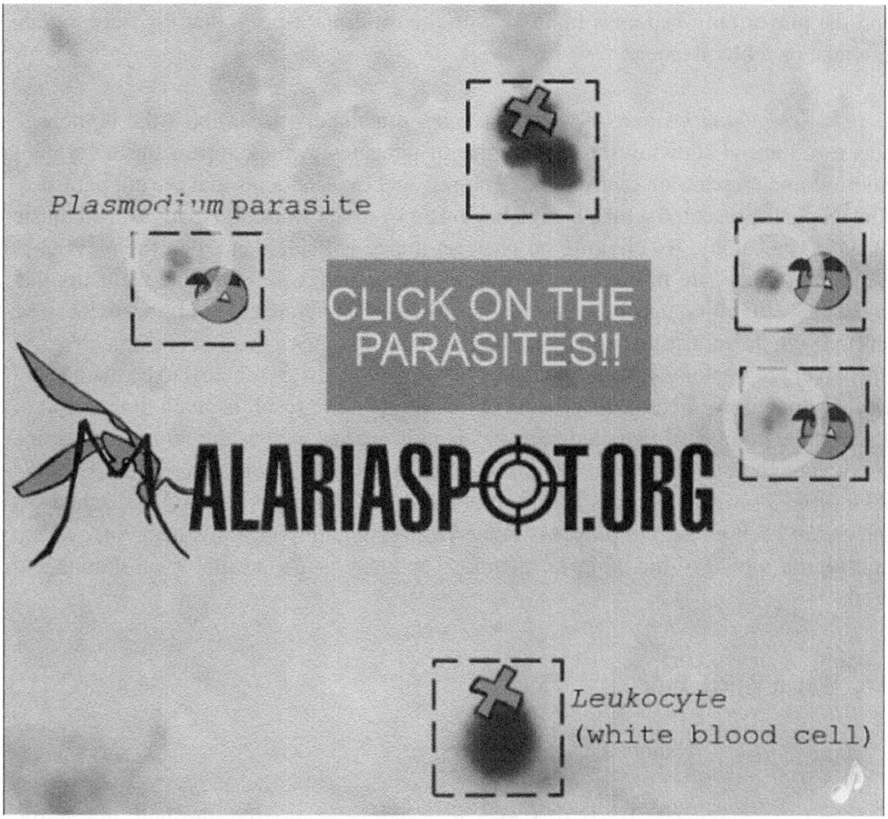

Figure 2.4. *Malaria Spot Game. The homescreen of the game acts as a tutorial: circles designate the forms corresponding to parasites to be discovered, while the crosses indicate cellular structures that must not be confused with a plasmodium*

For the designers, the results obtained validate the ludic approach for the counting of parasites on blood smears by crowdsourcing, by showing that non-experts are quickly able to identify the agent of malaria with an accuracy similar to that of trained technicians. The experience illustrates the potential of the ludic approach not only for the diagnosis and quantification of the malaria, but also more generally for the analysis of images for remote diagnosis purposes in the context of major public health problems.

Required human skills/player's education – The game only requires a sense of observation and speed, since a gamification element is to limit the time of observation to one minute. As for MALARIA TRAINING GAME, no prior knowledge is required and the player only becomes familiar with the different aspects that the *plasmodium* can take on a blood smear.

Challenge/gamification – The game is immediately accessible: the homepage acts as a tutorial showing three examples of parasites as they appear on the type of microscopic preparation that will be explored, and two elements that are not parasites. The whole is presented with a design borrowed from the world of *cartoons*, and music meant to be catchy. By clicking on *play*, an image is loaded and then the player has a minute to navigate through it and find all the parasites present. The elements that make the recognition task ludic are various: the whole picture cannot be seen, it can be shifted with the mouse through a black circle that simulates a microscope objective. As the player *clicks* along, a meter indicates how many parasites remains to be discovered while a countdown decreases, spicing up the search. The result is given in the form of a score, the number of detected parasites is indicated as well as the number of errors. If not enough parasites have been discovered, the note *you have discovered less than 90% of the parasites* appears and prompts to replay. However, *level 2* is reached. At the end of each game, the player is invited to log in and to send his/her score, and to consult the top 20 of the *malaria hunters*. The game is also available on iPhone and iPad.

2.6. Worm Watch Lab

Type of task – Nanotask.

Goal – WORM WATCH LAB[4] is a game accessible through the portal ZOONIVERSE. It is a project by the Medical Research Council (MRC) (Medical Research Foundation) launched in July 2013. The announced goal is a better understanding of the functioning of the brain and the influence of genes on behavior. The principle is that if a gene is involved in a visible behavior, then a mutation affecting that gene should lead to a significant change in behavior. The study model is the *Caenorhabditis elegans* nematode, classically used in biology, due to the speed of its development cycle, because its genome is fully sequenced, and that 40% of its genes have equivalents in humans. It is thus the opportunity to easily test the impact of certain mutations on behavior. The game consists of watching short video sequences showing a nematode, and in clicking on the *z* key whenever it is seen laying. As a matter of fact, egg-laying in the nematode is the result of a metabolic process, controlled by serotonin, a neurotransmitter also involved in the regulation of mood in humans. Thus, certain antidepressants, which stimulate the production of serotonin in humans also act on the nematode by stimulating its spawning. The identification in the worm of an increase or a decrease of spawning relative to the

4 http://www.wormwatchlab.org/.

normal can render a neural or muscular dysfunction with a genetic origin and be informative about the mechanisms behind depression in humans.

Benefits of crowdsourcing for the problem to be solved – Detecting worms that have abnormalities in the spawning rhythm is extremely long, because this implies observing thousands. As the MRC research team has also developed a microscopic observation system coupled to a computer that records the movement of the worm, it is extremely simple to visualize and report each laying. Since it does not require any skill or special knowledge, the idea of having it done by the followers of *citizen science* is at the origin of WORM WATCH LAB.

Required human skills – Almost nothing other than patience and *a contemplative character.*

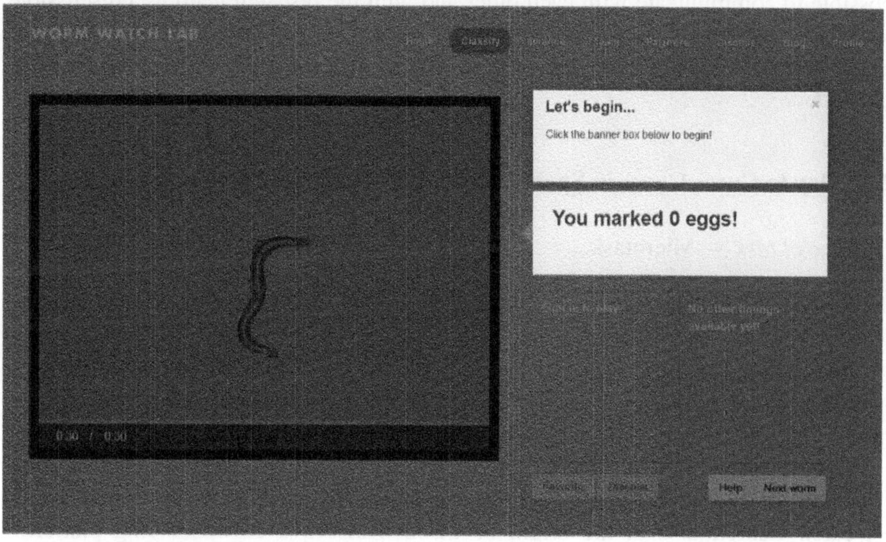

Figure 2.5. *Worm Watch Lab. On the left of the figure, the display allows visualizing a worm in motion for 30 s. The player clicks on the z key whenever it is seen laying an egg. On the right, the result of a sequence of observation is displayed, as well as the number of eggs observed by another player, if the sequence has already been viewed*

Challenge/gamification – Everything is done to make attractive and ludic the idea of... watching worms limply wiggle and expel an egg from time to time. But since this is not especially exciting, the designers are trying to encourage adhesion with motivation "Help researchers!", and with humour "Go hunting!". The interface is bright blue, the worm is represented in a *cartoonish* manner... By clicking on *Start*, a 30 s video starts. For each video, there is some brief information about the type of mutation affecting the worm found being observed, if it is known however. The *score* is equal to the *number of detected eggs multiplied by 10*. It is reset at each new video, but the profile saves the cumulative scores of the various observed videos, which are also stored in the profile in the form of icons. A 200 point bonus is given when the player is the first to watch a video. However, if another player has also viewed the video and reported spawning, his/her result is displayed, but does not appear to affect the score in case of difference in appreciation about the number of eggs laid.

Interactivity – Apart from the display of the number of eggs seen by the other player (when the video has already been viewed), there is no information about the other participants, no ranking, while it would be expected, as is the case in many GWAPs, to a validation by approval, that is to score points when the same number of eggs as any other player who watched the same sequence is declared. It is nonetheless possible to communicate with each other and with the research team in a discussion space, and notably to raise questions about a sequence in particular, either at the end of the observation by clicking on *discuss*, or from the *profile*, by clicking on the icon of a video.

2.7. Play to Cure: Genes in Space

Type of task – Microtask.

Game only available for certain models of smartphones - not tested.

Audience / popularity – Launched in February 2014, the game made it possible, in barely a month and a half, to achieve the analysis work that would have taken 6 months for a researcher. This represents more than 53,000 hours of gaming.

Goal – PLAY TO CURE is a game launched by the CRUK, the UK Cancer Research Center, also at the origin of CELL SLIDER, a crowdsourcing platform dedicated to the typing of tumor cells. The aim of the game is to to have gamers analyze data from DNA chips in order to highlight the altered areas of the human genome. At the controls of a spacecraft, the player travels the cosmos looking for *substance Alpha*. The route taken by the spacecraft based on the constraint imposed by the scenario will allow researchers to locate parts of the genome that may correspond to signatures of tumors. The ability to preferentially target certain regions of genome to identify areas potentially involved in any type of cancer represents a valuable saving of time.

Benefits of crowdsourcing – The main interest lies in the myriadization [SAG 11] of the task: the analysis of DNA chips is extremely long and spreading it to thousands of people represents a considerable saving of time.

Figure 2.6. *Play to Cure: Genes in Space*

Challenge / gamification – Although we were not able to test it (because we do not have a compatible smartphone), this game seems to be one of the rare *true* games, according to presentation videos and to the strong media coverage of its launch: the austere, monotonous and unrewarding task is completely hidden behind a science fiction scenario rendered by a rather convincing 3D design. The player at the controls of a spaceship is entrusted with a mission: collect the precious *substance Alpha* passing through the densest areas of the scenery, while avoiding and destroying asteroids. All the typical elements of a video game are present: levels to overcome, obstacles to avoid, shots, possibility of increasing the power of the ship, etc.

To boost participation, PLAY TO CURE largely invests on gamification, but without neglecting to strike on the sensitive chord: "Be a science hero! Play our new free mobile game to help our scientists to beat cancer sooner".

2.8. References

http://www.hfsp.org/frontier-science/frontier-science-matters/crowdsourcing-design-swarming-nanoparticles-cancer-applica.

https://www.ashoka.org/fellow/miguel-angel-luengo-oroz.

http://www.gamesindustry.biz/articles/2014-03-17-play-to-cure-has-already-analysed-6-months-worth-of-cancer-data.

http://scienceblog.cancerresearchuk.org/2013/03/01/can-the-power-of-the-public-help-personalise-cancer-treatment/.

3

GWAPs for Natural Language Processing

Since the computer revolution in the second half of the 20th Century, and particularly with the advent of the microprocessor in the 1970s, scientists have exploited the superiority of machines over humans concerning calculation speed. With regard to data processing, and in particular lexical data, pieces of software developed have clearly demonstrated the ease of access to information, which was not even imaginable with conventional processing. The problem of the acquisition of these data then arose, an even more critical problem that the necessary resources were increasingly voluminous. The advent of the Internet, which has allowed access to gradually more significant information databases, has caused this phenomenon to increase in an exponential manner. It was only at the beginning of the 21st Century, with the democratization of the Internet, that collaborative methods emerged. One of the most famous examples is the creation of the collaborative encyclopedia Wikipedia. Collaborators are fully aware that they contribute to the creation of a large-scale resource, in a *serious* manner. GWAPs constitute another way to address the collaborative creation of resources, where Internauts contribute to the creation of a resource while *playing*, usually even without suspecting it. In this chapter, we will present the delicate question of their acquisition after having explained the need for lexical resources in natural language processing (NLP), and then we will describe the main GWAPs dedicated to NLP, still active at the time of writing this book. Taking into account the evolution of this field, our list is not necessarily exhaustive, and we have only retained the most perennial GWAPs and those that have led to the creation of quality resources.

3.1. Why lexical resources?

Lexical resources constitute the main source of linguistic information for the automatic processing of natural language, NLP, where automatic text comprehension can be found at the forefront. Whether it concerns a *fine* comprehension (e.g. what is the meaning of this sentence? or even, does this sentence have a meaning?) or a *coarser* understanding (e.g. which field does this text relate to?), it is essential to have the possibility to access lexical resources in order to answer these questions. Globally, these resources are both about linguistic facts (*tomcat* is synonymous with *cat*) and the reality of the world (a *cat* has a *tail*). More accurately, among the tasks that require lexical resources, we can cite the determination of semantic proximity and similarity, the extraction of lexical information, the semantic annotation of documents, the development of paraphrases, semantic disambiguation, automatic translation, etc.

These lexical resources are mainly designed in the form of dictionaries that give the meaning of words or alternatively (and less frequently) in the form of lexical networks explaining the possible relationships between words. These two forms are not independent; we will see in Chapter 5, with the presentation of JEUXDEMOTS, that the lexical network JDM can take the form of a dictionary, called DIKO.

Whatever their forms, these are necessarily relatively voluminous resources: a traditional dictionary contains several tens of thousands of entries, Wikipedia contains approximately 1.5 million items in its French version, the lexical network of JEUXDEMOTS has more than 400,000 terms and 12 million relationships. We should note, however, that for "common" communication, a speaker only needs about 5,000 words.

It seems to us necessary here to draw the readers' attention to the availability of resources. Increasingly, more resources are freely available, and their owners authorize their use, or even their transformation or enrichment. A growing number of authors actually believe that open access is the best way available to developers of lexical resources to successfully face the four main challenges that are:

– visibility: a resource that is not visible will not be found by potential users and therefore will not be used;

– immediate availability: this is an indispensable condition for a resource to be used;

– perennial availability: freely available resources are more likely to be used in the long term, even when developers have stopped working on them;

– scalability: a resource that focuses on language, it must be able to evolve with it.

3.2. GWAPs for natural language processing

In the context of NLP, the first GWAPs emerged toward the middle of the first decade of the 21st Century. It is, therefore, a relatively recent domain, currently fast-developing, but not yet *stabilized*: as evidence, a number of GWAPs have had a reduced period of life, from a few months to a very small number of years, and the created resource has not been maintained; in addition, if it has not been given free access, it is generally no longer available.

3.2.1. *The problem of lexical resource acquisition*

The methods of acquisition of lexical resources have significantly evolved over the past few decades. Currently, they can be classified into five broad categories:

– Automatic acquisition from a corpus: very large corpora are currently available through the Web, and automatic processing is technologically possible in non-prohibitive times. Nonetheless, the result can be marred by background noise, hence the need for filtering, such as to retain only the significant words. The result can also be biased, either by the corpus itself (generalist corpora, such as newspaper articles, overlook knowledge that is basic or obvious to a human, and encyclopedic corpora do not usually mention news-related information) or by the extraction method (which can, for example, promote co-occurrences or misinterpret some assertions); in addition, all the information is not necessarily available in texts (intuitive and implied information). This process requires manual verification, at least on a sample of the acquired resource.

– Acquisition through unique or scarcely redundant contributions, remunerated or not. These contributions are often the result of the work of experts, long and costly, and the result is often static, hardly evolutionary (one of the most caricatural examples is the dictionary of the Académie française). In this case, manual verification by experts generally appears not indispensable. In addition, once the resource is *officialized* and made available, any correction can prove difficult (for example, most "paper" dictionaries have only a single yearly edition).

– Acquisition through highly redundant microcontributions from a large number of contributors: they are acquisition systems by a *myriadization of parceled work* (MPW) of which a critical analysis can be found in [SAG 11]. Contributors, often referred to as *turkers*, are generally (lowly) remunerated as in the Amazon Mechanical Turk service. This type of method is based on the fact that many Internet users are willing to collaborate on microtasks, most often at the price of a symbolic fee, sometimes even voluntarily. The large number of involved contributors will enable the development of large-sized resources. Unfortunately, experience shows that quality is not always as expected: some biases may be introduced by the decomposition into microtasks and it is difficult to obtain an acceptable quality from complex tasks. This mode of data acquisition is not limited to the field of NLP: the Amazon Web Services[1] site presents a version of Amazon Mechanical Turk as "a marketplace for work involving human intelligence". The given examples relate to photo/video processing (for example, the classification of objects in images), verification and cleaning of data (for example, the

1 http://aws.amazon.com/fr.

detection of duplicates in catalogs), collecting of various pieces of information (for example, respond to surveys), etc.

– Acquisition through contributing games, GWAPs, where players contribute to the acquisition of a resource, sometimes without being aware of it: this is a resource corresponding to common general knowledge, developed by a large number of players, hence the risk of errors, involuntarily introduced or not, which must imperatively be minimized. It is a relatively fast, inexpensive and evolving process. This approach is motivated by the observation that a group of individuals acting in an uncoordinated manner generally obtain better results than an isolated individual, even if he/she is an expert in the field. The number of potential players will be here all the larger so that the knowledge required for playing (the fundamental knowledge of natural language) is shared by the largest amount of players. In addition, collaborative systems, such as Wikipedia, have clearly shown the enthusiasm of Internauts for these projects. However, it will be necessary to verify the validity of the generated resource, but this verification can also be done using GWAPs (for example, TOTAKI in the project JEUXDEMOTS).

The border between unpaid MPW and GWAPs, sometimes somewhat blurred, is located only at the level of the ludic dimension of the latter; these two methods of resource acquisition are based on the phenomenon of myriadization (or folksonomy).

– Acquisition or enrichment of lexical resources in a language from resources already available in other languages. Various processes have been developed; they are mainly based on the hypothesis that the concepts and relationships between concepts are independent of language, a hypothesis which can introduce bias in the obtained resource.

3.2.2. *Lexical resources currently available*

The creation of lexical resources is closely linked to the history of writing. It goes back at least to the 3rd millennium BC, with the Mesopotamian tabular collections which constitute the first monolingual *dictionary*-type resources. In parallel, bilingual resources and multilingual resources emerged, allowing the translation and transmission of information among people. One of the most famous examples is the Rosetta Stone, which allowed Champollion to decipher hieroglyphs.

Two major technological revolutions, printing in the middle of the 15th Century and then computer technology in the second half of the 20th Century, will have a preponderant influence on how to design, develop and disseminate lexical resources. The first dictionaries worthy of the name emerged at the beginning of the 17th Century. The first edition of the dictionary of the Académie française was published in 1694, and "paper" dictionaries as we know them today, comprehensive and with detailed explanations, appeared during the 19th Century. An important work was published in 1852: this was *Roget's Thesaurus of English Words and Phrases*; it is based on the notion of concepts and thus allows access to words from their meaning. At the end of the 20th Century, most major dictionaries are computerized, which facilitates access and updating; their content evolves with the introduction of multimedia data; resource acquisition methods are diversifying, with the appearance of (semi)-automated methods and collaborative approaches (including GWAPs that are of interest to us here).

Electronic lexical resources appeared at the beginning of the 21st Century. These are the resources intended to be mainly used by machines. Lexical databases provide a structured and explicit description of entries (lexical units). The information provided is phonological, categorical (for example, grammatical category), quantitative (for example, frequency of occurrence in a corpus), lexical (for example, synonymy), etc. In parallel to these lexical databases (whose names often end by *-lex*), emerged lexico-semantic networks (whose names often end by *-net*). Each network node is a lexical unit (term or concept), and the relationships between these nodes express relations of lexical (synonymy, antonymy, etc.), ontological (hyperonymy, etc.) or another type (free association, associated feeling, etc.) of nature. Let us cite these main networks:

– WordNet (sometimes called the Princeton Wordnet) [MIL 90] and [FEL 98]: this is the most famous, its construction began in 1990 and it has more than 100,000 nodes, synsets, comparable to concepts each grouping a number of terms.

Following this, and sometimes based on its structure, other networks have emerged, including:

– EuroWordNet [VOS 98], then BalkaNet, SloWNet and AsianWordNet: equivalents of WordNet for European or Asian languages;

– BabelNet [NAV 10] which includes 3 million concepts and implements lexical and encyclopedic knowledge;

– HowNet [DON 06] of which each of the 1,500 concepts is characterized with a set of attributes and properties (lexical relations, semantic roles, etc.);

– WordNet Libre du Français (WOLF) [SAG 08], freely available, developed by automatic techniques.

Some projects have emerged independently from WordNet:

– Cyc (project started in 1984) [LEN 95]: semantic network of common sense based on logic, freedom of access, comprising 500,000 concepts;

– JEUXDEMOTS [LAF 07] lexico-semantic network, free to access, in permanent construction since 2007, comprising more than 420,000 terms connected by 12 million relationships; it is collaboratively built using GWAPs (see the presentation of the JDM project in Chapter 5).

Most of these networks can be the subject of some criticism. A number of them are not freely accessible, which seriously limits usage, as we saw at the beginning of this chapter. Furthermore, the relationships between terms are generally untyped (for example, the relationship between *cat* and *animal* seems analogous to that between *cat* and *mouse*) and rarely weighted (frequent relationships are then implemented analogously to unusual or rare relations). Finally, the determination of the meanings is either non-existent or simplified, or too complex, such as synsets in WordNet, which ultimately do not really correspond to the use of the speakers. Among the listed networks, only JEUXDEMOTS, which is freely accessible, is made up of typed and weighted relationships that can determine, thanks to the players, different uses for the same term (polysemy). These different meanings (or refinements) are also weighted, which allows the more commonly accepted meaning to be highlighted.

Toward what would be desirable

Without being utopian, or writing science fiction, the interpretation of texts as a human might do requires ideally access to a lexical and semantic network consisting of typed and weighted relationships between terms. This network should include both generalistic terms, representing a *common general culture* in order to concern matters that relate to everyday life, and specialized terms, labeled by this (or these) specialty (specialties), in order to be able to interpret texts from specific areas. Such a network should have at least several hundred thousand terms connected by a number of relationships in the order of a few million as a simple numeric comparison, the human brain has approximately 100 billion neurons. These relationships would originate from both knowledge of the world (for example, *a cat eats mash* or *hunts mice*) and lexical knowledge (for example, *cat* is synonymous with *puss*).

The weighting of a relationhip relative to another is difficult to quantify. What does it correspond to? To the *intrinsic strength* (difficult to define) of a relationship, or to its frequency (on what basis, or relative to which corpus)? In addition, the strength of a relationship may depend on the context. For example, it is easily conceived in a general manner that the cat–mouse relationship has a larger weight than the cat–allergy relationship, while in the medical field the second relationship is more preponderant. In JDM, the simple (simplistic?) solution which was adopted establishes the weighting by counting (a logarithmic function of the counting would probably be closer to human perception). It would also be possible to consider a weighting of the terms themselves, eventually depending on their frequency (in which corpus set?). Here also, this weighting would be dependent on the general or specialized concerned area.

Such a network makes possible the determination of the different meanings of a term, and in a probably finer manner than that encountered in traditional dictionaries, which may lead to a determination of the usages: the tree of the meanings has one or even two levels in traditional dictionaries, whereas it generally has several levels in the case of the lexical network JDM. There also, care should be taken, the determination of meanings can depend on the context; for example, the term *VTOP* usually means *voluntary termination of pregnancy*, with the exception of the aircraft domain where it takes the meaning of *vertical take-off plane*.

In summary, a lexical resource that would be desirable is a lexico-semantic network:

– in constant construction-evolution (to take account of developments in the language, discoveries, news, etc.);

– where the terms would be weighted by their frequency of use, but also tagged by their application domains;

– where the terms having multiple meanings or uses would be *refined* in as many subterms (for example, *cat* would be connected to *cat>animal* and *cat>discussion*);

– with typed, weighted and oriented relations; in addition to *classical* relationships (for example, synonymy, antonymy, hierarchy, etc.), it could include *less standard* relationships (for example, relationships related to emotions or judgment, various semantic roles, typicality relationships for locations, temporal values, etc.);

– which could also involve objects agglomerated with a relationship between a couple or a triplet and other terms (for example, the sentence "in tales, ogres eat small children with gluttony" constitutes an example where a couple of terms is connected to other words: *eat [subject] ogre* associated with a manner relationship to *with gluttony*, patient relation (object) to *children* or *small children* and typical location relationship to *tale*).

In fact, by reading the following chapters and notably the one dedicated to the project JEUXDEMOTS, it is correctly shown that the lexico-semantic network that is referred to by the latter is an example of a resource that seeks to meet these characteristics.

3.2.3. *Benefits of GWAPs in NLP*

The advantages of the use of GWAPs for the acquisition of lexical resources are multiple:

– NLP requires huge resources, probably hundreds of thousands (or even millions) of terms and tens (or even hundreds) of millions of relationships between these terms to represent not only the general common knowledge resources, but also the specialized resources to process specific areas: resorting to GWAPs would make it possible to create a generalist *universal* and multi-specialized network, which is nearly impossible to achieve with the other current techniques; the problem of the acquisition of resources (see section 3.2.1), common to other areas (see Chapters 1, 2 and 4), is therefore all the more crucial in NLP that the resources to be collected are particularly large.

– Unlike other acquisition methods, GWAPs make it possible to take into account the dynamic and evolving aspect of language and knowledge: actually, new words (for example, *H5N1 virus* and *folksonomy*) and new meanings for already existing words (e.g. *virus* and *tablet*) are constantly emerging.

– In addition, everything is not written in texts, in particular the information that is obvious for a human (for example, have you ever met a sentence advising you against the use of your smartphone in the shower?); it is, therefore, necessary to collect usage (and non-normative) resources, which are usually implicit in texts.

– Finally, GWAPs allow the incorporation of resources by consensus; natural languages are not part of exact sciences; the acquired resource will be all the more reliable if it is the result of consensus between several people, rather than emanating from a single one.

In the following, we present the main GWAPs dedicated, closely or not, to NLP and the acquisition of lexico-semantic resources.

3.3. PhraseDetectives

Type of task – Microtask, even though the task to achieve may be relatively complex.

Launch – December 2008.

The first version of PHRASEDETECTIVES[2] was launched online in December 2008. This game was designed by three British academics at the University of Essex: Jon Chamberlain, Massimo Poesio and Udo Kruschwitz [CHA 08].

Goal – Annotation of anaphoric relationships, which basically represents a relatively complex task.

What is it about? In grammar, an anaphora (also called grammatical anaphora, as opposed to rhetorical anaphora) is a word which, in a statement, provides a semantic repetition of a previous segment called antecedent. Therefore, there exists a relation between the antecedent and the anaphora: the anaphora is the element that represents, and the antecedent is the represented element. Thus, the anaphora enables avoiding repetitions. The first example given in PHRASEDETECTIVES is the following: "Sherlink Holmes went to the shop. He got some tobacco for his pipe". The word "He" is highlighted and it is indicated that it refers obviously to "Sherlink Holmes".

2 http://www.phrasedetectives.org.

PHRASEDETECTIVES thus comprises a training phase where the novice player learns the task that he/she will have to achieve, by means of examples made by experts. Then, the player must annotate some texts; if several players chose different annotations for the same text, it is presented to a larger number of players for validation. Unlike most GWAPs, here, the player is fully aware that he/she is contributing to a *serious* task: it is hardly fun, except for "lovers of literature, of grammar and of language" (as stated on the homepage of PHRASEDETECTIVES), to perform annotations on anaphoric relationships. The main *gaming* aspect of this GWAP lies in its presentation, with a little virtual character that can be likened to a detective and a system of points that enables a ranking between the players to be established. In addition, a Facebook version has been developed, allowing access through social networks.

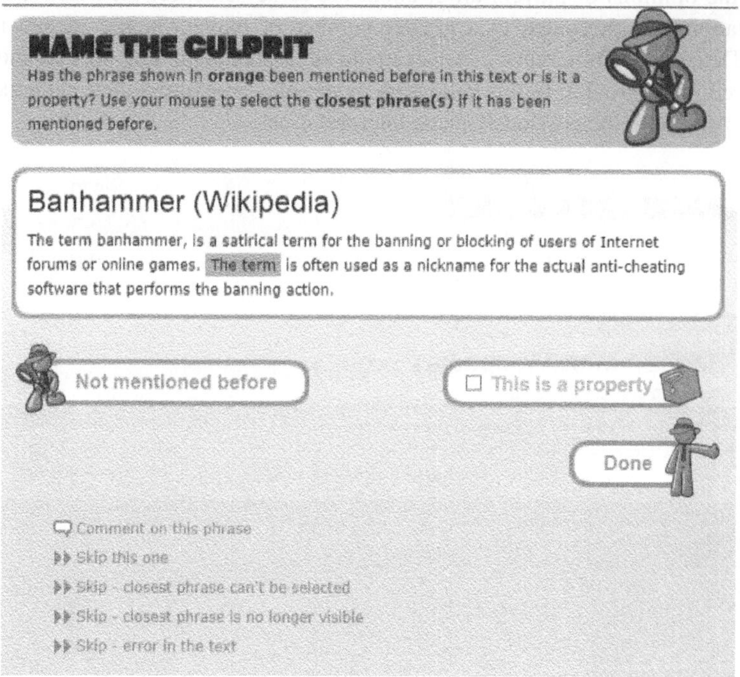

Figure 3.1. *PhraseDetectives: "Name the culprit". The player is asked what term (if it exists in the displayed text) constitutes the antecedent of the anaphora highlighted. When the player moves his/her mouse over the sentence, the terms or groups of terms become highlighted. The player also has the opportunity to state that it is not an anaphora, by clicking on "Not mentioned before" or "This is a property"*

In a more detailed manner, PHRASEDETECTIVES contains two game modes/levels: an annotation mode and a validation mode. Even before being able to access the first mode, the novice player follows an apprenticeship during which he/she is given a number of phrases for which the correct answers are known; in case of incorrect responses from the player, explanations are provided to help him/her understand his/her mistakes. When a player has given enough correct answers, he/she can access the first mode in which his/her answers will be saved. In the first mode, called "Name the Culprit", a text is presented to the player in a direct annotation mode, where an anaphora is highlighted. An example of it is given in Figure 3.1. The player must then indicate which word in the text constitutes the antecedent of the anaphora. The response of the player in the example of Figure 3.1 is presented in Figure 3.2. It may happen that the player estimates that the highlighted term is not an anaphora, in which case he/she can mention it by clicking on "Not mentioned before" or "This is a property": in the latter case, he/she will indicate of which term it is a property (in order to help players, examples of these various cases are given on the site). If the opinions of players differ about the same sentence, then their different interpretations are presented to a greater number of players in the second mode, called "Detectives Conference". An example of a screen of this second mode is reproduced in Figure 3.3. In this validation mode, players express their agreement or disagreement, with the previously made interpretations.

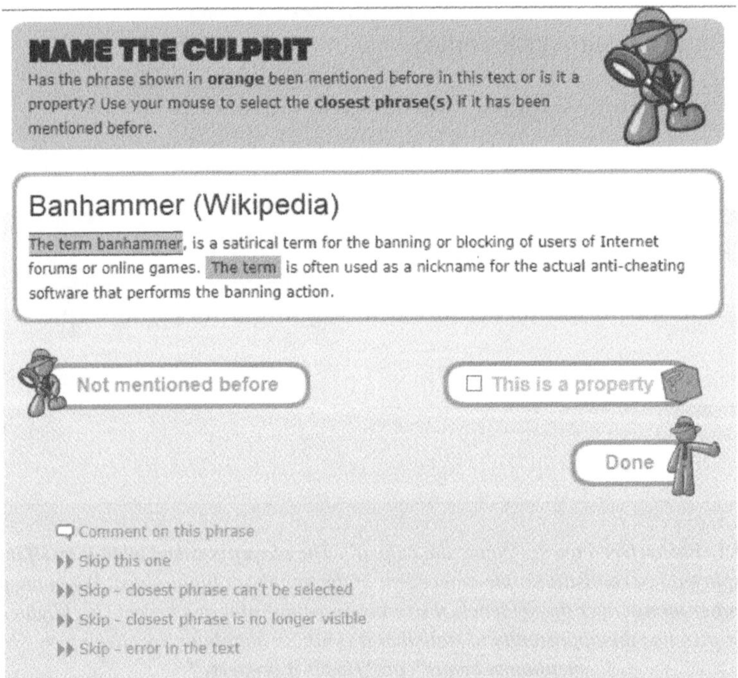

Figure 3.2. *PhraseDetectives: "Name the culprit". Screen showing the player's response (The term banhammer), highlighted. The last thing for him/her to do is to click on "Done"*

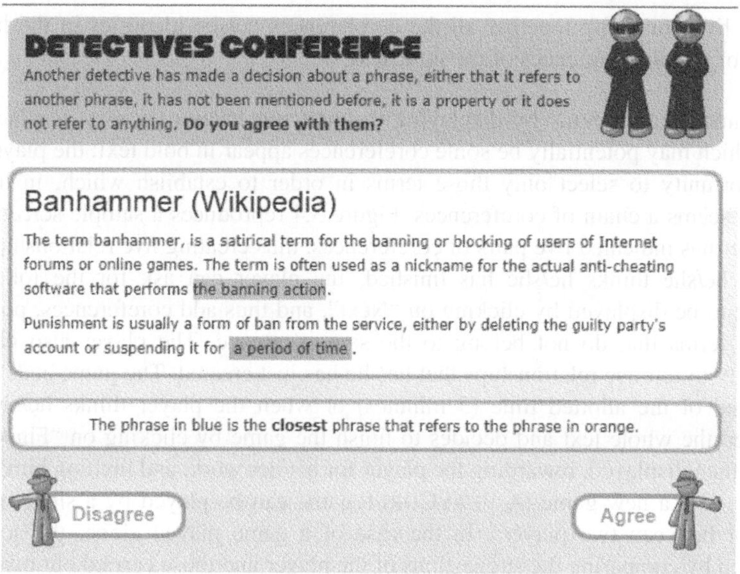

Figure 3.3. *PhraseDetectives: "Detectives Conference". The player is asked to validate or not, the proposal made earlier by another player. If the player disagrees with the proposal, he/she will have the possibility to indicate which term, according to him/her, is the antecedent of the anaphora, analogously to the mode "Name the Culprit"*

3.4. PlayCoref

Type of task – Microtask, even though the task to achieve may be relatively complex.

Launch – 2009.

PLAYCOREF[3] was developed at Charles University, in Prague, by Barbora Hladka and Jiri Mirovsky [HLA 09].

Goal – Annotation of coreferences in text data.

The term "coreference" is used when several terms in the same text refer to the same entity; in linguistics, a coreference is connected to its antecedent. For example, in the sentence "the kitten has died, it has fallen from the roof, it's like that, it has slipped on I don't know what", two of the three terms "it" (the first and the third terms) refer to the term "the kitten"; these three terms form a chain of coreferences. During a game, the player must, within a time limit, create as many relationships as

3 http://www.lgame.cz.

possible by connecting together all the coreferences he/she identifies in the largest number of possible sentences of the same text.

A game match begins by displaying the first sentences of a text in which the terms which may potentially be some coreferences appear in bold text; the player has the opportunity to select only those terms in order to establish which, in his/her opinion, seems a chain of coreferences. Figure 3.4 reproduces a sample screen after the player has indicated five pairs of coreferences, thus creating five relationships. As soon as he/she thinks he/she has finished, the player can ask for the following sentence to be displayed by clicking on "Next", and thus add coreferences, possibly between terms that do not belong to the same sentence. The player also has the possibility to remove relationships that he/she has just created. The game ends either at the end of the allotted time (5 minutes) or when the player thinks he/she has processed the whole text and decides to finish the game by clicking on "Finish". A score is then displayed, rewarding the player for his/her work, and inviting him/her to continue with a new game. A PLAYCOREF game can be played by a single player alone, or between two players. In the case of a game played alone, the score is calculated by comparing the suggestions of the player and those carried out manually or by an automatic procedure. In the case of a game between two players, the comparison between their suggestions is integrated into the calculation of the score.

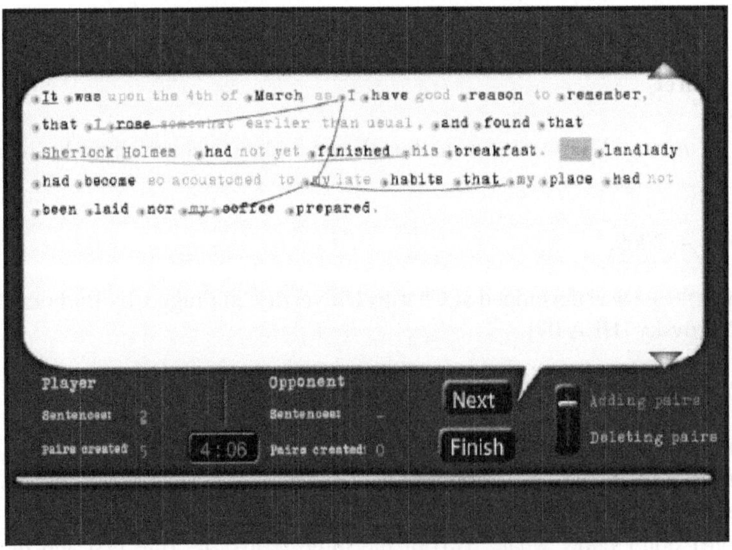

Figure 3.4. *PlayCoref. Example of a screen showing the coreference relations proposed by the player*

There may be noted a certain analogy in the desired goal between PLAYCOREF and PHRASEDETECTIVES (seen in the previous section). PHRASEDETECTIVES asks for the previous antecedent, whereas PLAYCOREF requests all the coreferences; in PHRASEDETECTIVES, it is possible to provide additional information (such as "Properties").

The analysis of the first tests with players, carried out by the authors of this GWAP, shows very encouraging results; in particular, some games between two players have allowed a larger number of coreferences to be detected than with manual annotation (were they all valid?).

Two other games are accessible from the portal of PLAYCOREF:

– SHANNON GAME: a sentence with missing words is displayed; the player has several attempts to guess them;

– PLACE THE SPACE: the player must insert *space* characters in a sentence where all the words are concatenated.

Are these two games really GWAPs? We do not believe they are.

3.5. Verbosity

Type of task – Microtask, even if a game is a form of dialogue between two players.

Launch – 2005.

This game was developed by Luis Von Ahn (one of the pioneers of the GWAPs), Mihir Kedia and Manuel Blum of Carnegie Mellon University [AHN 06b].

Goal – Collect *common sense facts*.

A common sense fact is a true assertion that is known by the vast majority of humans (for example, "fire burns" or "water soaks"). Such a basis of assertions is necessary in order to program *intelligent* systems. The number of common sense facts (estimated at hundreds of millions) is so important that conventional collecting methods have always proved ineffective, hence the idea of resorting to a GWAP. It is simultaneously played with two players: a *descriptor* player must coerce a *guesser* player into guessing a secret word, randomly determined by the system, by providing him/her with some clues; the roles are reversed for each game. The user interface offers input areas of semantically typed text that the *descriptor* player must give information about; the *guesser* player, therefore, sees the typed clues being displayed and can make suggestions. The first player can only input as clues terms that are known by the system. The semantic typing of the clues is based on sentences to be completed (for example, "the target term is a kind of...", or "the target term is used to..." or even "the target term has..."). Figure 3.5 reproduces an example of a game for a *guesser* player, while Figure 3.6 presents an example of a game for a *descriptor* player.

Figure 3.5. *Verbosity:* guesser *player. Here, the player must guess the word from the proposed clues*

One of the major problems is that there is no control whatsoever over the clues provided by a player, typically typed, but this typing is not necessarily respected; in practice, the game often sums up to giving a series of clue terms freely associated with the target term. Some kind of spontaneous discussion often begins between the two players. What is the reliability of the information thus saved? Hard to say, because the collected data do not seem free to access. It should be noted that in JDM, if typing is not respected, the intersection between the suggestions of the players is likely to be empty.

It would seem that the site GWAP.com from which it was possible to access VERBOSITY is no longer accessible (recently); in the blog at GWAP.com, it is written that its authors have migrated toward other projects.

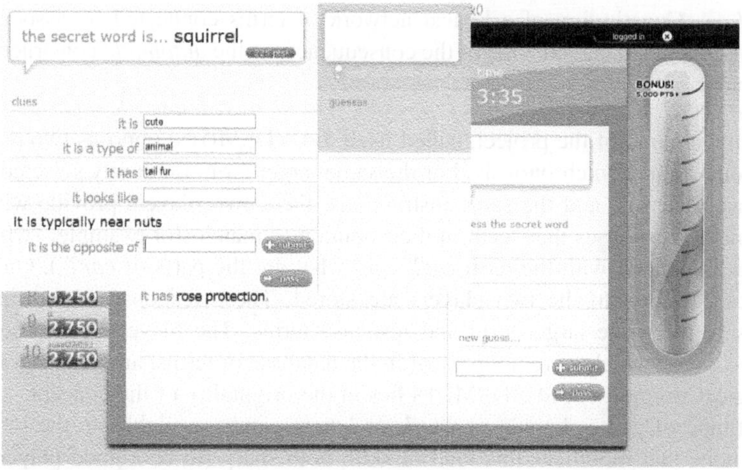

Figure 3.6. *Verbosity:* descriptor *player. Here, the player must make other players guess the word from the proposed clues. The clues proposed here follow the instructions of types, but no verification is carried out about this matter in real time. In addition, it is sometimes difficult to give information about some clues (for example, here: what is the opposite of "squirrel"?)*

3.6. JeuxDeMots

Type of task – Microtask, for most games of the project JEUXDEMOTS[4]. For the main game, one match lasts usually 1 min or more. It can then be considered as a succession of microtasks.

Launch – July 2007. Its first presentation was made by Lafourcade [LAF 07].

JEUXDEMOTS will be detailed in Chapter 5: it features different games and counter-games. The latter make it possible to both offer diversity to players and to verify the data collected through the main game. The evaluation of the acquired resource is also carried out using TOTAKI, a GWAP in which the player can sometimes *feel more clever than the machine*. In addition, a small number of players have expressed their wish to get more involved by participating more *accurately* in the construction of the network: the latter have, therefore, the possibility of *contributing* through the DIKO interface, that is to directly give information, for each term of the network on the relation(s) of their choice. These contributions will be manually validated by the game administration.

4 http://www.jeuxdemots.org.

Goal – Constitution of a lexical network of terms connected by associations typed, weighted and tagged. One of the consequences is the *automatic* construction of a dictionary.

The main game of the project, called itself JEUXDEMOTS, opposes two players, anonymously and asynchronously. For the same target term, randomly selected from a database of terms, and the same instructions, these two players provide within a time limit the responses that seem in their opinion to match (for example, "what are the ideas associated with the term *cat*?" or "what are the parts of *car*?"). Only the responses common to the two players are stored. These responses determine the relations between the target term and response terms. The players' responses can introduce words which were not present in the database of terms, and thus can enrich it. The great strength of JEUXDEMOTS lies in the originality of the concept: among the multitude of games known as *words* or *letters* games available on the Internet, there is none that mobilizes this kind of skill, both sharp (to be a good player, it is necessary to have an extensive vocabulary and/or be quick and organized to provide a maximum of answers in the allotted time) and universally shared (it is about providing general knowledge conveyed by a common support: language). Briefly, it is a game accessible to all, in which it is possible to progress and improve quickly by observing and imitating the strategies of the best players. The developers of the project are striving to maintain the excitement caused by the discovery of the game by regularly adding either new games or variations in the main game so as to avoid boredom by proposing a maximum of diversity. Since its launch in 2007, thousands of players connected, participating in total to more than 1.4 million matches only for the main game. The result is the creation, from a coarse base of 150,000 of isolated words, of a lexico-semantic network, which (as of October 2014) counts more than 400,000 terms connected by 12 millions of relations.

3.7. Zombilingo

Type of task – Microtask.

Launch – 2014.

ZOMBILINGO[5] is one of the most recent GWAPs to appear on the Net, developed by Karën Fort and Bruno Guillaume of the University of Lorraine and in [FOR 14].

Goal – Dependency syntax annotation of corpora.

Each sentence added in the base is automatically preannotated (subject, verb, complements, etc.) using syntax analyzers. The players are then asked to indicate

5 http://zombilingo.loria.fr.

their annotation for this sentence. If a number of players considered as sufficient give an opinion contrary to the automatic preannotation, the annotation of the sentence is then changed. Since completely annotating a sentence can be a fairly complex task, ZOMBILINGO breaks down the overall task into a series of more elementary tasks in order not to discourage the player. The latter chooses a type of task (annotation of the subject, object, attribute of the object, etc.), follows the corresponding training on a reference corpus and then can start playing.

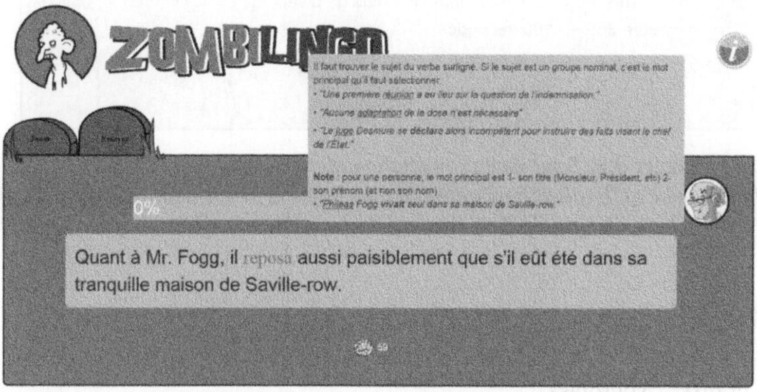

Figure 3.7. *Zombilingo: annotation of the subject. The player must indicate which word of the sentence (As for Mr. Fogg, he rested as peacefully as if he had been in his quiet house in Saville Row) is, according to him/her, the subject of the verb (rested)*

The interface of the game, the world of zombies, has been especially studied to attract players and encourage them to play in order to generate sufficient annotations. An example of a screen, concerning the annotation of the subject of a sentence, is reproduced in Figure 3.7. A second example, concerning the annotation of the object of a verb, is reproduced in Figure 3.8. The generated data are freely accessible from the website. This game is clearly aimed at people who maintain a good memory of grammar at school, and who consider logical analysis as an entertaining puzzle. It is nevertheless regrettable that the wording of the tasks to achieve is somewhat opaque: "P obj agt","dep" or even "Affixe" absolutely do not provide any information about what to do. Suddenly, the player may be tempted to click randomly in the proposed first sentence so that the immediate correction that is then displayed comes to clarify things a little. Maybe this is desired; perhaps it is wanted that the player discovers by himself/herself, by trial and error, what is expected of him/her, etc., or is it a *defect of youth* of this GWAP, still in its launch phase; it is safe to say that in this case it will be quickly corrected. However, for people who have some knowledge of grammar, it is a little bit annoying that one has to err to discover to which grammatical concept must respond to the required annotation.

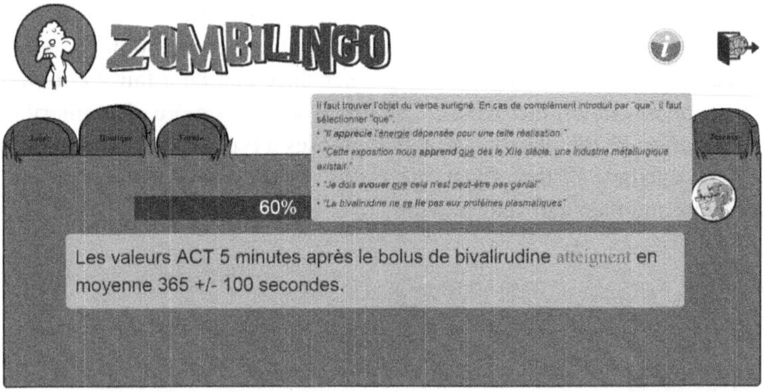

Figure 3.8. *Zombilingo: annotation of the complement of object*
In the sentence ACT values 5 minutes after the bolus of bivalirudin reach in average 365 $\pm$ 100 seconds. the player must annotate the complement of object of the verb reach

3.8. Infection

Type of task – Microtask.

Launch – 2014.

INFECTION[6] is a relatively non-standard game since it is a Video-GWAP to the extent where the user does not provide a textual response, but behaves as in an action game, by shooting fictional characters (simplified, it must be said). This game was developed by Daniele Vannella and Roberto Navigli of the University of Rome [VAN 14].

Goal – Validation of associations between concepts.

In the scenario of INFECTION, humanity is attacked by a virus: part of the population remains healthy, and another part is infected. All the humans move toward a town *a priori* not yet infected. The player must stop infected humans to prevent the propagation of the epidemic, but leave those who are healthy. Since all humans look alike, the game is based on a question-answer password system to distinguish them: the responses of healthy humans are consistent with the question, and those of infected humans are not (for example, if the password is “medicines”, a healthy human can answer “radiology” or “drug”, while an infected human will give a response unrelated to this topic).

6 http://knowledgeforge.org/.

The mission of the player consists of leaving alone humans who seem to him/her to be responding with a term in connection with the *pass theme*, and stopping others. The game is divided into several levels of difficulty. Players obtain points based on their ability to save humanity; a general table showing the scores of the players displays where they stand and increases motivation. By modifying the password question, INFECTION allows free association, synonymy and antonymy relations to be obtained. The players are supposed to play honestly, but a simple mechanism controlling the responses of players has been implemented to detect malicious infringements.

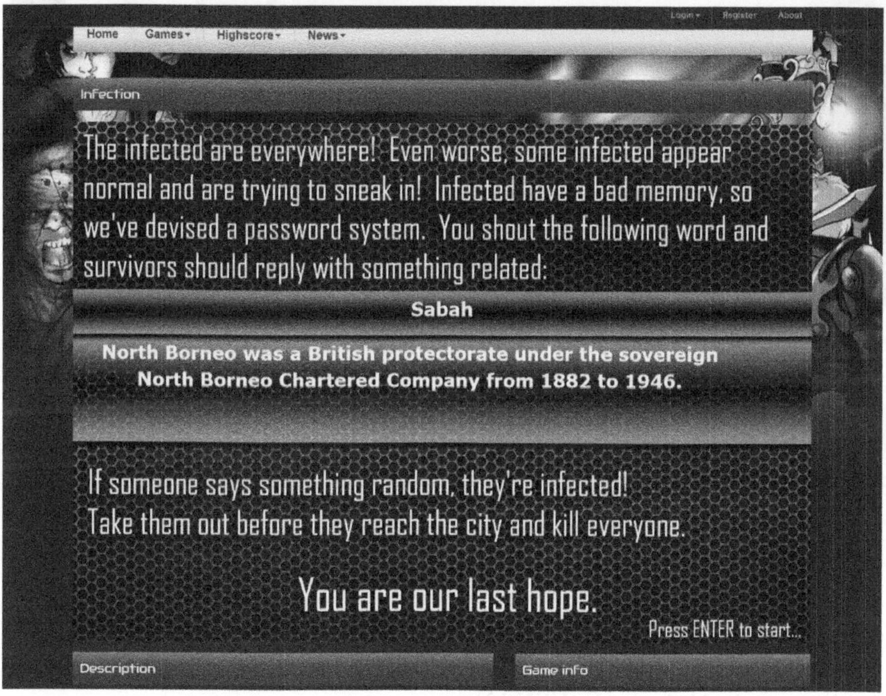

Figure 3.9. *Infection: instructions*

It should be noted here that Jurgens and Navigli [JUR 14] developed an original approach, to the extent that, although these games relate to lexical annotation or disambiguation, no action of textual nature is required on the part of the player, the game mode being closer to that of *shoot 'em up*-type games.

Figure 3.10. *Infection: the launched match*

3.9. Wordrobe

Type of task – Microtask.

Launch – 2012.

WORDROBE[7] was developed by Johan Bos and his team at the University of Groningen in the Netherlands [VEN 13].

Goal – Enrich the Groningen Meaning Bank (GMB), a freely accessible semantically annotated corpus.

At the beginning, WORDROBE comprised a single game, *Sense*, designed to collect lexical disambiguation data in order to contribute to the construction of the GMB, a valuable reference tool for the NLP. The principle is that of multiple choice questions (MCQs): a sentence comprising a word in bold is displayed, and the player must choose the appropriate meaning of the related word from a list of suggestions. A match

7 http://www.wordrobe.org.

comprises about 10 sentences to be annotated in this manner, and the number of points earned depends on the degree of similarity of the responses with those of other players for the same sentences. As the games are stored, the score varies even when one does not play, since the played matches are confronted with those of others. A cursor makes it possible to place a bet on the answers which spices up the game, since a significant bet is a double-edged sword: the potential gain is not only proportional to the bet, but it is also the potential loss. One can try to show off.

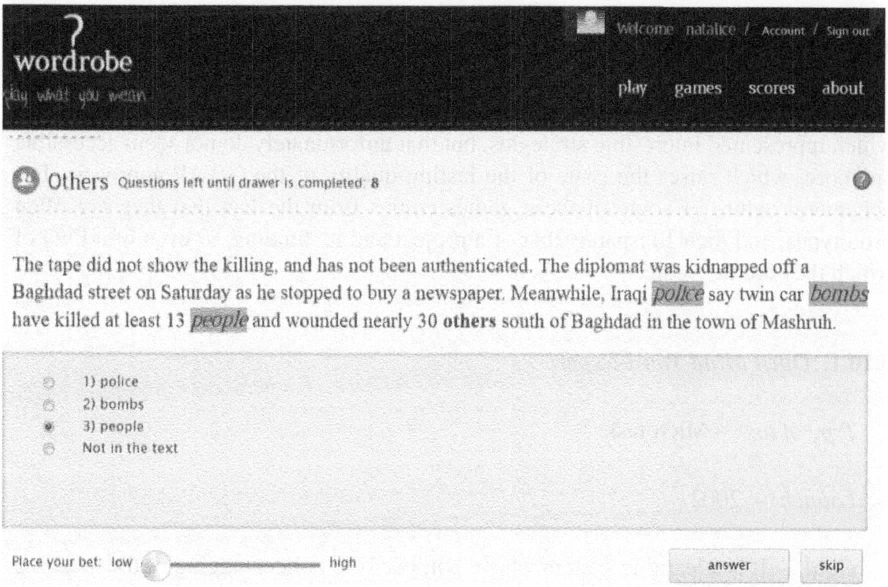

Figure 3.11. WORDROBE*: a game from* Others*, one of the seven annotation games In the displayed sentence, the player must indicate to what word highlighted corresponds the word* **others** *displayed in bold, by selecting the corresponding item in the suggested list below. The cursor at the bottom left allows bets to be made on one's response, and one's gain to be maximized (but taking the risk to obtain the opposite effect, if the answer given is not that chosen by the majority of the other players)*

WORDROBE has quickly enriched itself with new games, and today represents a series of seven different annotation games, all based on the principle described above. As a result, in the proposed sentence, the goal may be to choose the proper antecedent for a pronoun (*Pointers*), or to identify the *kind of entity* that a proper noun designates (*Names*). It may also be about assigning an *agent*, a *cause* or an *instrument* status to a term in connection with an action verb (*Roles*), or to say whether the indicated term refers to a human, an animal, a concrete object, an abstract entity, etc. (*Animals*)

Since the difficulty is variable, all the proposed tasks can satisfy every audience, "from amateur linguists to language gurus" as the slogan on the home page claims. Like many games with a lexico-semantic goal, one must still be a passionate lover of language and the subtleties of vocabulary to become a very productive and hardworking player. Despite the relative variety of tasks, the principle of the game is always the same, and ludic elements (bet, classification and various awards) fail to make this annotation activity exciting in the long term.

3.10. Other GWAPs dedicated to NLP

In the following, we briefly present a number of other GWAPs dedicated to NLP, which represented interesting strategies, but that unfortunately do not seem accessible anymore, which raises the issue of the lasting quality of the GWAP approach. The ephemeral nature of some of these games comes from the fact that they are often prototypes, and their lifespan is that of a project and its funding, or even of a PhD of which they are the subject.

3.10.1. *Open Mind Word Expert*

Type of task – Microtask

Launch – 2002.

Goal – It is a learning system whose purpose is to collect tagging of the meaning of polysemic words in a corpus.

OPEN MIND WORD EXPERT was developed by Timothy Chklovski of MIT and Rada Mihalcea of the University of Texas [MIH 03] in the early 2000s. In this game, several sentences comprising a polysemic word are suggested to the player, as well as the various meanings of this word extracted from WordNet. For each displayed occurrence of the polysemic word, the player must indicate what sense among those presented appears to be the most adequate; he/she also has the possibility of clicking on "unclear" or "none of the above".

Due to its principle of resource acquisition, OPEN MIND WORD EXPERT is one of the *ancestors* of GWAP; but given its interface (which is by no means ludic), can it be actually regarded as a game? As proof, its authors when presenting it speak of users or contributors, but never of players. We would speak instead of an acquisition system by MPW.

3.10.2. *1001 Paraphrases*

Type of task – Microtask.

Launch – 2005.

1001 PARAPHRASES was one of the very first GWAPs, developed by Timothy Chklovski [CHK 05b].

Goal – Collection of data allowing machine translation systems to better recognize the different variations of paraphrase.

During a match, players are asked to indicate paraphrases for an expression that is given to them (for example, "this can help you"). If the player indicates a paraphrase already previously provided by another player, he/she wins a certain number of points; otherwise, his/her paraphrase is added to the one already proposed, awaiting future matches. This was probably the first GWAP dedicated to NLP that uses the validation method by pairs of players (pairs or more, in the case of 1001 PARAPHRASES).

3.10.3. *Categorilla/Categodzilla*

Type of task – Microtask.

Launch – 2008.

Goal – Collection of associations between terms and their characteristics.

This GWAP was developed by David Vickrey's team at Stanford University [VIC 08]. In CATEGORILLA, players are asked to provide terms (words or expressions) corresponding to the provided criteria (for example, "specify objects that fly") and starting with a particular letter (for example, the letter "B"; and in this case, the answers can be "bird" or "Boeing"). Each match takes place between two players, who receive points when their answers are consistent (as in JDM). Players can enter as many answers as they wish, and at the end of each game all the answers are displayed. This game has allowed for a significant number of hyperonymy/hyponymy-like relations to be obtained, or providing the typical semantic roles of agent/patient (relation linking an action to its typical patient or agent). CATEGODZILLA is a slightly different version: the constraint corresponding to the first letter is either deleted or subdivided between easy and difficult letters.

3.10.4. *FreeAssociation*

Type of task – Microtask.

Goal – Collection of associations between terms.

In this game, also developed by David Vickrey at Stanford, players are asked to provide words in relation to an indicated target term, without being allowed to give *forbidden* terms from a *taboo* list. The principle of this GWAP resembles that of JDM, by prohibiting the players from using the words more strongly connected to the target term, and therefore promoting the emergence of less common relations, belonging to the long tail[8] (the many infrequent relations).

3.10.5. *Entity Discovery*

Type of task – Microtask.

Launch – 2010.

Goal – Collection of named entities.

This game was developed by Nathan Green's team [GRE 10] at the University of North Carolina. In this game, the player must indicate the types of named entities that he/she detects in the proposed sentence; he/she has the choice between "person", "location" and "organization". In practice, after a learning phase (where responses to the proposed sentences are known), the player is presented with phrases for which the system does not yet have answers. Each game is played between two players, anonymously, in a synchronous manner; this makes possible to indicate to each player, at the end of the game, the concordances between his/her answers and those of the other player, which determine the number of points obtained.

A variant of this GWAP, called NAME THAT ENTITY, is played asynchronously: the second player performs a form of verification of the answers given by the first player.

3.10.6. *PhraTris*

Type of task – Microtask, even if the requested task is not always *simplistic*.

Goal – Determination of the order of the terms in a sentence.

8 http://en.wikipedia.org/wiki/Longtail.

This is a game of syntactic annotation, developed by Giuseppe Attardi's team at the University of Pisa [ATT 10]. In this game, players must rearrange sentences in a logical manner, such as bricks must be arranged in the famous game Tetris.

A number of GWAPs have briefly emerged, but nevertheless some deserve to be mentioned:

– COMMON CONSENSUS, developed by Henry Lieberman's team at MIT, was officially presented for the first time in 2007 in [LIE 07]. It is intended to collect common sense knowledge about the goals which motivate humans in their actions of everyday life, as well as the means implemented to achieve this. The main reason that led to the development of this game is the note expressed by its authors about the fact that computers do not have a large amount of basic information about human behavior and environment (for example, "humans drink when they are thirsty" or "the sky is blue during the day"). It is estimated that the quantity of these knowledge elements amounts to billions, hence the idea of developing a GWAP to collect them.

The principle of COMMON CONSENSUS is analogous to that of a TV game show (*Family Feud*) where two families compete by answering simple and open questions and where their responses will be compared with the responses of the community. In the televised game, the responses of the community are predefined, developed by polls; in the case of COMMON CONSENSUS, the responses of the community are dynamically defined based on the responses of the players. During a game of COMMON CONSENSUS, a goal is initially randomly selected in a pre-existing base of goals (for example, "watch a movie"), then the raised questions are based on a *grid* of questions related to this purpose (for example, "why would you..." or "what objects do you use to..."). Some answers to these questions make it possible to determine new questions to ask in following matches, thus building a hierarchy among the goals. During a game, limited in time, players can chose as many answers as they wish. Each chosen answer is compared with all the answers already provided by other players on this subject; the player gets an immediate response in the form of a bar with a size proportional to the number of players who provided this same response. An adequate program enables two responses to be compared in order to determine whether or not they are semantically similar.

It would seem that COMMON CONSENSUS is no longer accessible.

– TRAIN ROBOTS is an annotation game similar to PHRASEDETECTIVES. In this game, the player has two images of a configuration of colored blocks: one representing an initial situation and the other representing a final situation. A robotic arm is also represented on the screen. The player must provide, in natural language, the series of operations that will allow the robotic arm to achieve the final configuration from the initial configuration. The game thus achieves the collection of lexical data that will enable programming *intelligent* robotic systems.

– ONTOGALAXY is a game developed by Markus Krause and his team at the University of Bremen. Its objective is the acquisition of common sense knowledge of words, in particular to complete ontologies. The interface of this Video-GWAP represents a space battle scene; the player at the controls of a spaceship must shoot objects, according to given instructions. The latter is given in terms of gaming, and not in terms of resource acquisition (for example, "shoot all the vessels representing objects that can be touched" means "what are the objects that have the property of being concrete?"). By doing so, the player associates target objects with source terms and predicates (for example, "in the displayed image, what are the objects larger than a car?"), or objects with a given property (for example, "in the displayed image, which objects are hot?").

– FACTORY GAME: in this game, an assertion is displayed; the user can then indicate whether, in his/her opinion, it is true, false, if he/she does not know, or if it does not make any sense (for example, "electrical wires are placed in houses without living room"). The response provided by the player is compared with the majority response of the other players; the number of points he/she obtains depends on the result of this comparison.

Other letter and word games

Many other letter and word games exist on the Web; some are similar to GWAPs because they store the answers of the players when they are different from those expected by the system. However, a few of them are actually GWAPs insofar as there is no verification of the data (which may be wrong in this case) provided by the players, and therefore no valid learning is performed by the system: the games played do not result in the creation of a quality resource.

Among these games, we can cite the following:

– GUESSWHAT?: it also concerns a guessing game, relatively simple. The player must guess a word corresponding to the hints displayed on the screen (e.g.: "animal with soft fur, with sharp fangs, whiskers and usually a long tail") whose number of letters is indicated (e.g.: "3 letters"). The player's response is recorded by the system if it is different from that expected. The player also has the option to quit if he/she does not find anything. This is not a GWAP, no really valid resource being created by the players.

– FIND THE LINK: the software displays two words, or more, characteristic of the word to be found; the player has a choice between several suggestions. Unlike JEUXDEMOTS, it is a closed game, it is not the player who makes the association suggestions and there is no learning by the machine.

4

Unclassifiable GWAPs

This *holdall* chapter, which is obviously not exhaustive, shows the great diversity of studies and/or research areas likely to give rise to games with a purpose (GWAP). The diversity comes with a relative disparity regarding the popularity of the games, which itself directly reflects the interest they generate through the mass of potential contributors that the general public represents. It should be noted that it is easier to raise feelings and civic pride with projects whose benefits are on the medical level, than by reaching out to support researchers in quantum physics, or to participate in more or less tedious classification tasks of different kinds. To compensate in the areas where adhesion is related neither to affection nor to the empathy generated by major public health issues, GWAP must motivate potential contributors by *pulling on other strings*. In other words, the pleasure provided by gaming must be larger so that the value related to the *greatness* of the cause to support is modest.

4.1. Beat the Bots

Type of task – Neither microtask nor macrotask.

Goal/presentation – The designer of the game is the company ShareThinkLtd that is behind VOUCHSAFE[1], a system dedicated to the fight against spammer robots. VOUCHSAFE is a ludic alternative to the system of CAPTCHAS, is pleasant to use and is more effective. The principle is as follows: instead of typing on the keyboard a series of more or less legible letters, the user who wants to be identified as a human will have to solve a simple visual puzzle.

1 http://www.vouchsafe.com/.

The puzzles can be of two types:

– find the two objects that have the highest number of relations between themselves: faced with five or six images of everyday objects that simultaneously appear on the screen, the user will have to connect with a stroke the object that is assigned to them in the textual instruction to the one that is the most related at the functional or the thematic level;

– find the intruder: among the five or six presented objects, it is necessary to designate the one that has no connection with the others by surrounding it with a stroke.

BEAT THE BOTS[2] is the first game prototype intended, on the one hand, to test the technology VOUCHSAFE and, on the other hand, to gather data in order to train the software about the way of reasoning of humans in order to make it more effective in distinguishing humans and robots.

In addition, the designers are protesting against the use of conventional crowdsourcing to achieve large-scale or complex tasks. According to them, it is a method that gives good results but that is questionable from an ethical point of view. They rather advocate the ludic approach, which they name *crowdplay*[1].

Benefits of crowdsourcing for the problem to be solved – VOUCHSAFE technology exploits the fact that the human brain differs radically from a machine in its way of perceiving visual information and to associate meaning with it. Nothing is as effective as making a human answer... to test a system aimed to distinguish between human responses and those automatically generated by the software. The associations that players spontaneously make between the displayed objects are used to help the latter to develop new visual challenges, more intuitive and more *human*.

Required human skills – None, because the point is precisely to be human and to react with spontaneity and speed; this last quality only is useful for the *gaming* aspect.

Interactivity – FRAXINUS, BEAT THE BOTS is accessible via Facebook and, therefore, makes use of all the infrastructure of the social network to promote its own distribution by encouraging players to contact their *friends*. In this game, one plays against the machine and not against another human; therefore, the interactivity lies in inciting the players to compare their score with that of their *friends*, and in expressing themselves about the game by joining a dedicated Facebook page.

Challenge/gamification – The game is fun and it is easy to get carried away to make several games in a row. The task, seemingly simple, is spiced up by the time

2 https://apps.facebook.com/beatthebots/.

limitation: a match comprises 15 groups of images and only 2 min are available to process the whole. (It is possible to *pass* a series of images against a penalty (loss of points)). Thus, there is no time to hesitate, which encourages spontaneous responses. At the end of the game, several classification parameters are communicated: the number of points, which depends on the number of correct answers, as well as the response time, the quickest response time for the series, the number of correct answers and the series of correct answers. The goal is to make a good score to get into the *hall of fame*, which displays the top 100 players. In addition, the top 5 and the top 15 players can, respectively, display *gold* and *silver trophies* on their wall.

Design – The design is simple, colorful, a little flashy, and is accompanied by a somewhat pounding music and sound effects that fortunately can be deactivated. Players immediately know what to do; a game lasts only for 2 min; it is therefore expected to quickly complete one game without having to delve into the rules and complicated strategic choices, or to play several games depending on the time available. One of the great advantages (at the gameplay level, and consequently at the at the level of the utilization as robot/human discrimination procedure of the robot/human discrimination procedure) suggested by the designers is as follows: as playing requires nothing else than to draw a line or make a circle, the game and the finished product are particularly suitable to the touch screens of smartphones and tablets.

4.2. Apetopia

Type of task – Difficult to classify according to microtask/macrotask criteria.

Goal – APETOPIA[3] is a game which was launched by the University of Berlin, and is intended to provide data on how the shades of color are perceived by people in order to model the best color parameters.

Benefits of crowdsourcing/required skills – The interest is directly related to the high variability with which the human brain discriminates shades of color, which will allow the development of a large database. No special cognitive skills are required, except for the gaming aspect: the same dexterity and quickness as those necessary in video games (such as *running games*), are required.

Challenge/gamification/design – The gamification is very extensive; the player can completely forget that they are is collaborating in the creation of scientific data. The interface resembles that of any video game: a series of passages with two gates scroll on the screen. By navigating with the keyboard keys, the player has to go

3 http://colors.htw-berlin.de/.

through the doors whose color is the closest to that of the sky by avoiding various obstacles (which cause *life points* to be lost) and by collecting as much *gold coins* (bonus) as possible. The series of doors accelerates as the game progresses, making the challenge increasingly difficult. The game does not require registration; the player at the end can choose to communicate, or not, his/her score to be registered in a daily and global ranking.

COMMENT.– The game aspect is perfectly convincing, from the scenario, ergonomics and challenge point of view. APETOPIA is an easy game, accessible to all the public, which requires no special skills except manual dexterity that all the experts of this kind of game usually possess. It is difficult to go further into the analysis because there is no detailed explanation about the scientific purpose, the relevance of the results achieved relatively to the goal and the way in which they are employed. Too bad. Would this be an abandoned project?

Figure 4.1. *Apetopia. The home screen of the game*

4.3. Quantum Moves

Type of task – Macrotask.

Goal – The game was launched in beta version in 2012 through the portal SCIENCE AT HOME[4] designed by the Department of Physics and Astronomy of the

4 http://www.scienceathome.org/.

University of Aarhus in Denmark. Between the time when it was tested and the drafting of these pages, it was interrupted. Its upcoming reactivation with an improved version is announced on the portal. The game is based on the simulation of atomic movements that can be obtained in laboratory using a laser beam. In the game, the mouse movements are actually simulating the laser beam and the objective is to move an atom into the target area in an optimal manner and within the shortest time. The solutions of the players who have achieved the highest score are compared with those derived from computation algorithms. They are then tested and optimized in laboratory to identify which can be used in the solution of complex problems in quantum physics. In the long run, the aim is to help researchers in physics to build a quantum computer to perform combinatorial computations that cannot currently be achieved by conventional computers.

Benefits of crowdsourcing/required skills – At each level of the game, the challenge is to beat the score of the computer, which means to achieve a more efficient transition than that predicted by computation algorithms to change an electron from one state to another. The only quality required is manual skills. QUANTUM MOVES[5] is a game of dexterity.

Challenge/gamification – The game is currently unavailable according to the home page; an improved version is being designed. As far as we remember, the difficulty gradually increases, which is strategically well thought. This is an effective manner to hold the interest of the player who easily overcomes the first levels, which consist of a tutorial; as a result, the player will want to continue.

4.4. Duolingo

Type of task – Rather microtask.

COMMENT.– In the world of GWAPs, and among those analyzed here, DUOLINGO[6] stands out from the others insofar as it is a business for profit-making the data provided by players are not used to make knowledge progress, to help researchers and to cooperate in solving complex problems, but constitute a service billed to customers in the potentially very important market of translation.

Audience/popularity – Launched in June 2012 on the Internet, DUOLINGO is developed for iPod, iPad and iPhone from November 2012, before being available for Android smartphones in May 2013. On 14 October 2013, DUOLINGO claimed 10 million users and then advertized its partnership with online news websites CNN and

5 http://www.scienceathome.org/index.php/game/play-beta.

6 https://www.duolingo.com/.

BuzzFeed for the translation of their articles. This partnership is presented as a means of maintaining free of charge the online language learning site. The fact that Duolingo has been elected *Application of the Year 2013* by Apple has caused an increase in the number of users from 16 to 20 million, and the designer confirms that the application attracts now 100,000 users more each day.

Goal – DUOLINGO is a platform launched by Luis Von Ahn, the father of CAPTCHAS and the concept of GWAP, which offers free language learning (English only, for French speakers, and several European languages for English speakers). Learning is made ludic, and the aim is twofold: data collected by crowdsourcing are analyzed in order to improve pedagogy and users are invited to test their progress by translating *real* documents, currently originating from BuzzFeed and CNN information websites. Some sources, by referring to Duolingo as "the application that really makes you learn a language and makes you addicted, while earning money on your back", do not hesitate to mention disguised volunteer work, a criticism already found elsewhere [GOO 13] for the initial discovery of Luis Von Ahn (CAPTCHA and reCAPTCHA). We have tested DUOLINGO for learning English.

Benefits of crowdsourcing for the problem to be solved – If the problem to be solved is considered as avoiding paying fees to professional translators, the benefit of crowdsourcing as employed by DUOLINGO is obvious: by distributing as exercises, the same fragments of text to be translated to thousands of users, and then by the synthesis of their outcomes, it is guaranteed that a translation of a quite correct level will be obtained free of charges.

However, the application also aims to be a tool for reflection on the didactics of foreign languages and uses crowdsourcing to test and permanently compare teaching methods in order to offer users that which has proved to be most effective. Finally, for the founder, making the teaching of a foreign language free to everyone is a philanthropic act, and the profit from translation sales only contributes to maintain and guarantee the application free.

Required human skills/player's education – In order to produce effective data, it is essential that the player should be assiduous and highly motivated by learning a language. According to the designer, achieving the totality of the proposed progression in English, for example, is equivalent to achieving level B2 of the *Common European Framework of Reference for Languages* (CEFR), or concretely to be able to understand what it is heard, to manage a conversation, to be able to understand an article or a novel in English, or even to watch a movie in the original language.

Interactivity – The player/student is encouraged (or even coerced) to invite friends and to compete against them. On the home page, several buttons encourage the use of social networks (Facebook, Google+ and Twitter) to attract contacts, to seek acquaintances among people already registered and to send e-mail invitations. A debate list is also accessible and it seems quite active. In the menu, a section called *activity* summarizes all of the actions of the player (e.g. has completed a unit, has translated x phrases, is now friend with such …). For each enumerated activity, other players can *like* or *comment*. Interactivity between players is also greatly promoted at the level of translation activities, since inside the translation module, one is invited to give an opinion about the translations of the other players, and the player only progresses as *translator* because of the positive appreciations of the others.

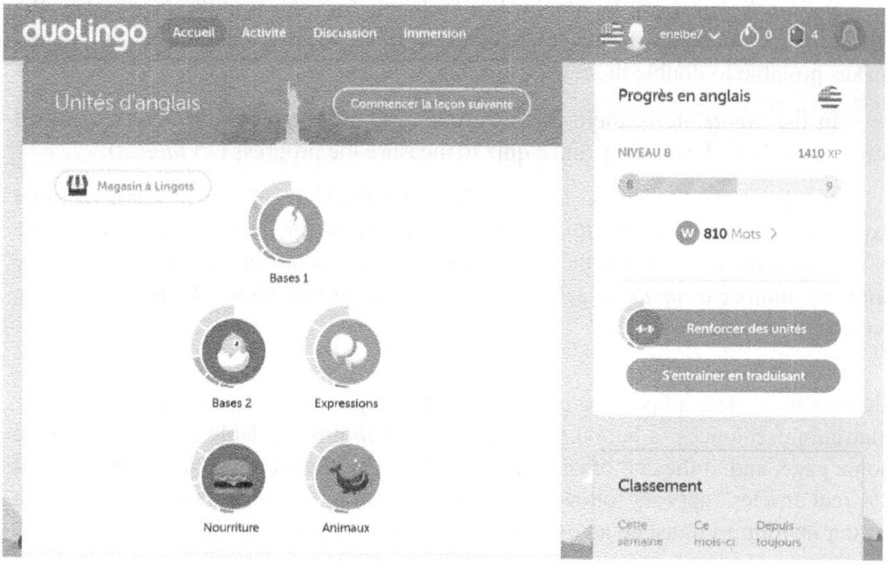

Figure 4.2. *Duolingo. The home screen of the game which summarizes the progress through the different learning modules, and offers either to strengthen knowledge or to test it by translating texts extracts from the press*

Challenge/gamification – DUOLINGO mixes learning and translation in a ludic interface.

Learning is organized into small *modules* or thematic units consisting of 1–10 *lessons*, each *lesson* consisting of a series of 20 questions. The ludic and motivating aspect is linked to several elements:

– the virtual currency of DUOLINGO is the *lingot*: the player earns *lingots* by accessing the next level, by completing a unit, by completing a *lesson* without any error, by inviting a friend and by being assiduous (i.e. by playing at least 10 consecutive days). Earned *lingots* enable making purchases and various transactions in the *ingot store*;

– the player has by default three *hearts* (equivalent to three life points) at the beginning of each *lesson*. At each error, he/she loses a *heart*, and a player who loses his/her three *hearts* is forced to repeat the series of exercises from the beginning (with other exercises). However, it is possible to buy a "spare" *heart* in the *lingots store*;

– each successful *lesson* is greeted by a brass band sound, and access is allowed to the next *lesson*;

– various bonuses can be obtained in exchange for *lingots* in the *Ingot store*: the *streak freeze* which allows a day of inactivity without penalty, a double or nothing that makes possible to double the rewards in *lingots* in a series of 7 days;

– in the *ingots store*, there are also trainings, unlockable with *lingots*: a timed training module (10 *lingots*), and a quiz to measure the progress (25 *lingots*);

– finally, with 30 *lingots*, *bonus skills* are unlocked: *idioms* to learn pictorial expressions in English, and *flirting* to learn sentences for flirting in English. There is no actual ranking, but on the home page, a progress bar increases proportionally with the number of *modules* achieved, and a score in Experience Points (XP) and a score in words are displayed.

COMMENT.– The player is constantly encouraged to test his/her learning by translating sentences: a button *training by translating* is available at any time on the home page, and at the end of each series of exercises the mention "now you can read x% real articles" appears followed by "do you want to see what happens?" and by a button inviting to translate an excerpt of newspaper article. The player can also click on *immersion* and select an article to translate. He/she is invited to approve/disapprove/modify the sentences already translated by other players, or to initiate a new translation, which will be submitted to the appreciation of the other players. Every player begins at the translator's *level 1* and the access to the next level is subject to obtaining at least 100 positive votes from others.

4.5. The ARTigo portal

Goal – The name ARTIGO[7] designates the portal of a project destined to facilitate navigation and searching in the databases related to reproductions of works of art, which can contain millions. The purpose of the games available on this website is to associate keywords with reproductions of works of art, in order to facilitate their search. The reproductions are those of the ARTEMIS database of the

7 https://www.artigo.org/.

Institute of History of Art of the University of Munich, which contains more than 25,000 of them.

Benefits of crowdsourcing for the task at hand – Assigning keywords to each reproduction is the only way to search and find a piece of art, a drawing or a painting. In addition, given the size of databases, the choice of relevant keywords for this recognition task is a prohibitive work in terms of time and personnel, and a task yet impossible to automate, which justifies the use of crowdsourcing. It is even more relevant that the assignment of keywords does not require particular expertise, and that as in many annotation games, only words proposed by at least two people without consultation are retained. The games are directly inspired by the ESP GAME by Luis Von Ahn [AHN 04].

Human skills required for the task at hand – The choice of appropriate keywords does not require any expertise or knowledge in art history, but still assumes an interest and a curiosity toward art with a good observation capability.

The ARTIGO portal has developed four games that are all aimed at the association of keywords to works of art through modalities that slightly differ. In order to inspire motivation, the designers clearly focus on the combination of ludic (they have made efforts to add excitement to the games) and cultural (works of art are discovered) aspects, as well as on the valuation related to the prestige of participating in a collective description of works of art.

4.5.1. *ARTigo and ARTigo Taboo*

Type of task – Microtask.

Principle – For ARTIGO[8], a work of art is simultaneously shown to two persons who must, each separately, and within a time limit (60 s), define the keywords that characterize, as precisely as possible, what they see, and not only concerning the iconography, but also the style, the period of time and the feelings that emerge from it. Points can be earned for each keyword also proposed by the other active player or by other players during previous sessions. In the TABOO version, the principle is the same but the use of certain words, already allocated, is forbidden [WIE 13]. These principles are not unlike those encountered in the various games of the JeuxDeMots project (discussed in Chapter 5).

Gamification – The game aspect is first linked to surprise: the works of art to annotate are randomly selected and presented without any thematic relation, neither

8 https://www.artigo.org/taggingGame.html.

of time nor nature. The time limit of 60 s to type the keywords that spontaneously emerge (or not) spices up the exercise, and the fact of seeing the words also suggested by the other player appear in blue is exciting. A game or *session* comprises five successive pictures and therefore, does not last any longer than 5 min. The simplicity and speed make it possible to play several successively, or complete one quickly without having to reconsult complicated rules. The only frustrating aspect is not being able to visualize, at the end of the game, for example, all of the keywords suggested by the other player, especially when there are none in common. At the end of a game (of a series of five works), a variety of information is available for each work (title, author, date and place), as well as the score that was achieved with this image and the summary of the provided words and words in common. By clicking on *to the table of honor*, a score is available cleverly divided into *best players of the day/best players of the month/best players of last month*. After a few games, there are a few chances not only to appear in the first or even in the second, but also to win places, which is stimulating and incites to replay.

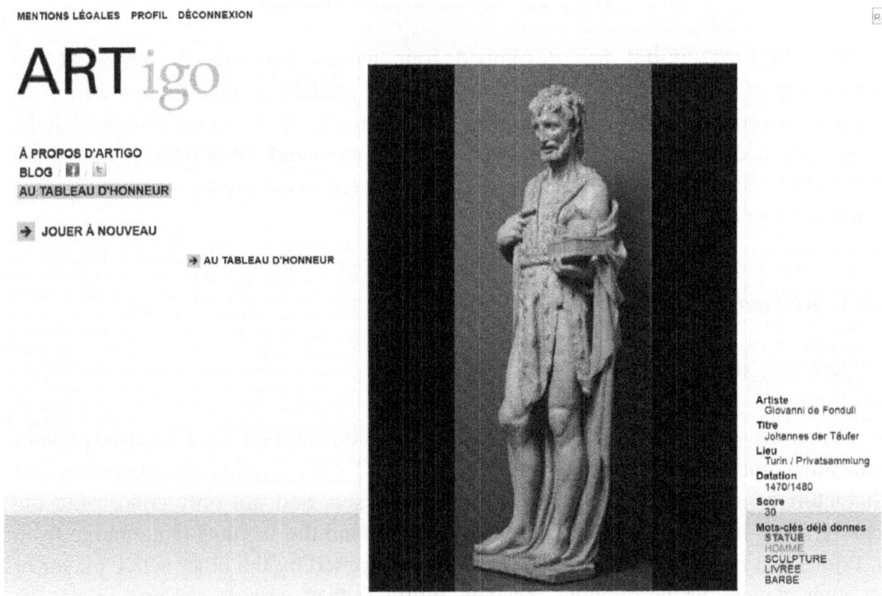

Figure 4.3. *Artigo. At the end of the game and for each work of art, the score that the player obtained with each is accessed (common words are displayed in another color), and so is the identification information (title, author, etc.)*

Interactivity – Interactivity is a key element of the game, since a game takes place simultaneously with another partner, whose number of suggestions can be seen increase in real time, and identical suggestions appear in another color. Points can be won in the event of identical answers. Responses are also confronted with those given for this work of art during previous sessions, and in the case of shared answer(s), it/they is/are displayed in pale color and also bring(s) points, but less. A blog is also available in German.

4.5.2. *Combino*

Type of task – Microtask.

Principle – A reproduction of a piece of art is proposed, together with a set of keywords relating thereto. The player has 90 s to associate two-by-two keywords chosen by successive *clicks* on the proposed words. The suggested words are keywords originating from the games ARTIGO and ARTIGO TABOO, and therefore words validated as relevant individual keywords because they were proposed by at least a pair of players.

Challenge/gamification – More choices are available by clicking on *more keywords*, and as for ARTIGO, points are earned as soon as the player suggests combinations identical to those of the opponent (here, the game is in beta version, the opponent is a computer program (bot)), whose number of suggested combinations can be seen increase in real time. However, less points are also earned if the combination has already been proposed by a human player during a previous session. The player can at any time move to another work, particularly when lacking inspiration.

The authors explain they have largely drawn their inspiration from ESP GAME [AHN 04], either for the design of the game or for the fine study of gamification elements aimed to boost the motivation of the player [STÖ 12, WIE 13].

4.5.3. *Karido*

Type of task – Microtask.

Principle – The purpose is similar to that of previous games, namely to assign keywords, with a slight difference though: the point is to constitute groups of keywords associating generic words and more specific terms. The principle is different since the two players do not have the same role. Nine reproductions of works of art are simultaneously presented to the two players: one describes a work by using keywords and the other must guess which one it is among the nine, and so on until all the reproductions have been processed. Thus, the game allows the validation of those words as keywords which, individually, make it possible to identify an image within a group. However, the main purpose of the game is to encourage players, by providing them images containing points in common, to find more specific keywords in order to allow a finer discrimination between images thematically or visually similar.

Challenge/gamification/interactiviy – Interactivity is really the driving force of the game since everything happens as if two players had to collaborate effectively to earn points and be strategic in a role as in the other. The game starts with the choice of the game mode (*time limit* or *turn limit*) which may also influence the outcome. As soon as this choice is made, nine reproductions are displayed and a role is assigned to the player, either *descriptor* or *guesser*. The *descriptor* is the one who chooses an image and types a keyword. The *guesser* must then click on the image to which he/she associates it. If it is the correct one, it disappears from the interface and the *descriptor* describes another. In the case of error, the *guesser* can ask a question, and the *descriptor* can specify another keyword. When the nine reproductions are recognized, the roles are reversed for a new series of nine images. KARIDO[9] is by far the most funniest game of the portal: the challenge which consists of giving the correct keyword so that the partner can immediately find what image it refers to is exciting. Limitations (in time or in the number of possible proposals) spice up the task, the need to be fast and accurate is stimulating and the player must be strategic in the choice (1) of the order in which images are described (keeping the most difficult to describe for the end is a guarantee of success) and (2) of the keyword, which must be the most accurate and relevant as possible in the case of images that are similar (for example, avoid the keyword *landscape* if among the reproductions there is more than one).

In fact, the human player alternatively plays the roles of *descriptor* and *guesser*, but in both cases he/she plays against the computer. In the role of *descriptor*, the program reuses the keywords defined by a human during a previous session, in the same order and within the same time frame. Additionally, in the role of *guesser*, he/she compares the keywords provided by the player with the keywords already attributed to the image in the database and guesses the image that is most appropriate by successive probability calculations [STE 11]. It should be noted that if both players were human, the game could be strongly biased by the nature of the clues given in keywords (for example, the *descriptor* player might give indications about the position of the image on the screen, and not on the image itself).

9 https://www.artigo.org/karido.

Figure 4.4. *Be A Martian. The home screen offers various crowdsourcing activities related to the collection and analysis of the mass of data collected by satellites and exploration rovers*

4.6. Be A Martian

Type of task – Microtask.

Goal – The exploration program of the planet Mars, conducted by the National Aeronautics and Space Administration (NASA), includes among others three satellites in orbit around Mars and two rovers, Opportunity and Spirit, which generate a tremendous amount of data to analyze. The NASA has, therefore, developed BE A MARTIAN[10], an interactive online space through which the public is invited to participate in three types of tasks:

– map Mars: the objective is to align pictures representing fragments of the surface on a more general map in order to achieve a precise mapping of the surface of the planet;

– count craters: on satellite pictures, it is necessary to locate the craters by clicking on them, in order to provide information about the relative age of different zones of the planet, which is proportional to the number of craters;

– tag Mars: the point is to characterize as precisely as possible the pictures taken by the rovers that travel on the surface of Mars, giving information about the aspect of the ground (rocky, sandy, etc.) and the sky (uniform, cloudy and starred) to classify the database of images of Mars.

10 http://beamartian.jpl.nasa.gov/welcome.

Benefits of crowdsourcing – The interest of crowdsourcing is both related to the huge amount of data to be processed and the difficulty to automate the fine characterization of visible elements in the images. The requested task is typically microtask-based, and does not require any particular expertise.

Challenge/gamification/interactivity – BE A MARTIAN is typically a crowdsourcing activity that tries to appear as a game by borrowing multiple ideas from the world of gaming. The scenery mimics the piloting center of a spaceship, where activities are proposed through control screens. On registration, the player chooses an *avatar* and a *pseudo*, and the various contributions achieved in the three areas give rise to titles, awards, badges, points and several prizes.

There is really not much interactivity, the player is alone, but lots of scientific information about the planet Mars and the NASA program are provided through the *Citizenship Hall*. The latter hosts crowdsourcing tasks and other popularization activities more or less gamified (quiz, tourism on the planet Mars, sending post cards to the exploration rover Spirit, etc.).

4.7. Akinator, the genie of the Web

Type of task – Microtask.

Launching – 2007.

This game was originally designed by Arnaud Mégret, as an amateur in 2007, but AKINATOR[11] only became really famous from 2009 due to its launching on smartphones. Figure 4.5 reproduces the home screen.

Goal – It is primarily a game, hence its development on smartphones. It is designed to collect the decisive/discriminant features of characters, and since a recent extension, of any type of entity.

A game takes place between the *genie* AKINATOR and a player. It is a guessing game where the player thinks of a character, real or fictional, and AKINATOR will attempt to guess his/her name by asking questions. The database of the game comprises approximately 100,000 characters with their characteristics; it selects the questions in such a way that it can eliminate at each player's response the greatest number of possibilities (for example, "is the character alive?", "is this a man?" and "is he European?"). The responses from the players are necessarily: *Yes, No, Do not know, Probably* and *Probably not*. The questions are organized and asked in a logical manner, as would a human do (for example, after a negative response to the question

11 http://fr.akinator.com.

"is he European?", AKINATOR may ask if the character is American). After some 20 questions, AKINATOR's base usually contains only one character corresponding to the responses of the player; the latter may then make a suggestion. If it is incorrect, it proposes the player to continue by asking a new set of questions (which are often designed to somewhat expand the search, assuming that the player could have been wrong a few times in his/her answers). If after these additional questions AKINATOR is wrong, or cannot find the character, it asks the player to tell him what was the character, and what characterizes him: it thus enriches its knowledge base. Thus, the game is based on the honesty of players, but it should be emphasized that it is not really motivating to make someone guess a character with deliberately false clues. What significance is there in providing such clues to make AKINATOR lose (the *trolling* rate should actually be very low)? The player responds mostly by Yes or No, but he/she has the opportunity to confess his/her ignorance, or his/her uncertainty concerning the question raised about the character he/she thinks. Figure 4.6 reproduces some screens of a game.

Figure 4.5. *Akinator: main part of the home screen. This screen shows a particularly ludic interface: the game is available in several languages (currently 14), it is possible to play on different types of smartphones, and there is a specific version for children. The results of the last 10 games are also displayed, thus showing* AKINATOR*'s success rate (90% on the screen presented here), on characters sometimes little known or little publicized*

Figure 4.6. *Akinator screens. Series of some images screens showing examples of questions raised by the* genie AKINATOR. *Observe its different facial expressions, in particular his satisfaction when he believes he has found*

AKINATOR displays on its home screen (Figure 4.5) the results of the last 10 matches: usually, its success rate is about 9 or 10/10. Today, more than 200 million matches have been played, and every day several tens of thousands of games are played in France, mostly on smartphones. It should be admitted that, first and foremost, it is a fun game, the GWAP side being reduced to those cases where AKINATOR was not aware of the character to which the player was thinking.

A new version of AKINATOR makes it possible to guess objects, animals or concepts, but most players prefer to make him guess characters. There is also a AKINATOR KIDS version intended for children, available in GWAP mode (such as the conventional version), as well as in educational game mode where AKINATOR provides clues to make the young players guess famous characters.

However, we might legitimately wonder if AKINATOR is actually a GWAP or a simple guessing game. The collected data are not freely available and their exact form as well as what would be possible to accomplish with them remains unknown to the general public. On AKINATOR's Website, the game is rather a showcase for the algorithms and the techniques that it uses than for its benefit in a possible collection of data.

4.8. References

http://fr.wikipedia.org/wiki/Duolingo.

http://www.slate.fr/story/83419/duolingo-appli-langue-rend-accro.

http://en.wikipedia.org/wiki/CommonEuropeanFrameworkofReferenceforLanguages.

5

The JEUXDEMOTS Project – GWAPs and Words

The objective of the JEUXDEMOTS project[1], described for the first time in [LAF 07] and then in much more detail in [LAF 10], is to build a large lexical-semantic network for the French language. It involves a collection of games and counter-games that together contribute in one way or another to this objective.

The presentation and the analysis of the games of the JEUXDEMOTS project can highlight the qualities that a GWAP must have, in particular the criteria that must be met to be effective. Moreover, the fact that the project was initiated in 2007 gives the necessary distance to assess its actual impact in terms of quantity and quality of the collected resources. In this chapter we discuss the limitations and biases of each of the games of the project and show how counter-games can be used to validate the data and to compensate, at least partially, the bias induced by the main game.

5.1. Building a lexical network

Lexical networks, whether they are general or specialized, are a particularly valuable resource for linguistic computing applications, notably in lexical disambiguation [VÉR 90], especially since the emergence of WordNet [MIL 90] and EuroWordNet [VOS 98], among others. In most of the available lexical networks, relations are not weighted; they are simply enumerated, without any indication regarding their *strength* or frequency. Weighting can be used to distinguish between strong relations and those that are more anecdotal, the latter being not necessarily less significant in natural language processing (NLP) processes. Propagation algorithms in disambiguation can take advantage of such weights. However, the

1 http://www.jeuxdemots.org.

question of the factors that determine the weight of a relation remains complex: it is strongly linked to the frequency of usage of the relation, but it is not the only factor. This type of resource can be manually constructed with more or less luck and/or by crossing data with more or less automated procedures, followed by a verification, also manually [SAG 08].

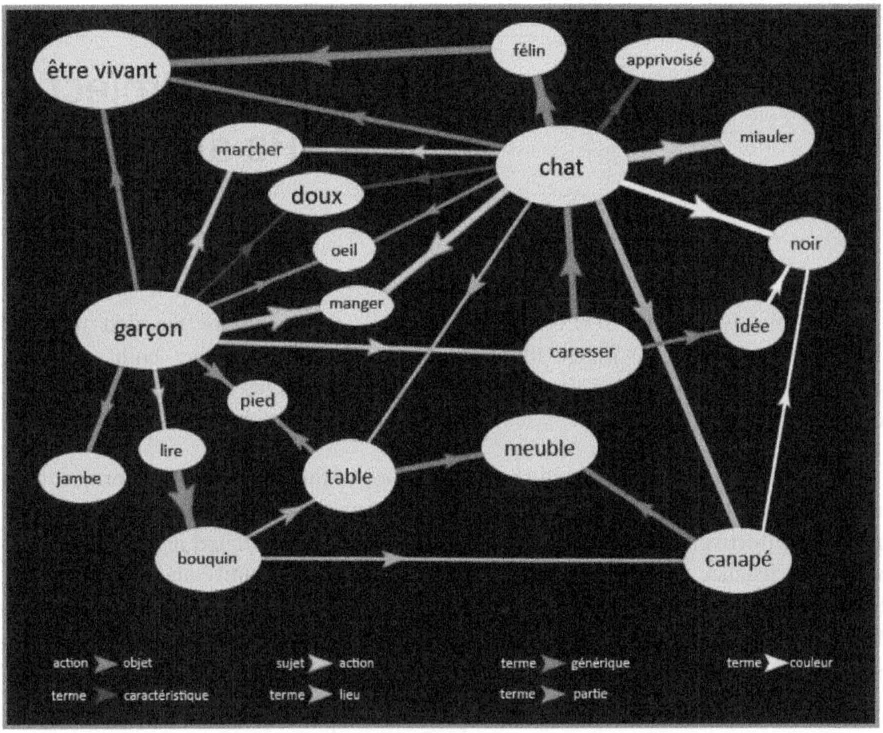

Figure 5.1. *Illustration of a lexical network. Diagram of a portion of a lexical network bringing forward some relations*

Moreover, lexical-semantic networks often contain a variable number of relation types, typically ontological (hypernyms) or lexical (synonyms and antonyms). HowNet [DON 06] is an example of such a lexical resource (Chinese and English) built manually and presenting many types of relations. Cyc [LEN 95] concretizes a major effort of manual construction of a database of general knowledge based on logic, whose form itself can be assimilated to a semantic network, but which is not lexical or scarcely lexical. Finally, some networks also contain monolingual or multilingual associative relations. For example, BabelNet [NAV 10] is a large multilingual network originating from the automatic extraction/compilation of information from Wikipedia.

5.2. JEUXDEMOTS: an association game

JDM, the first game of the project, was launched in July 2007 to build a lexical network for the French language. This network began with about 150,000 terms without any relation between them. After 7 years and 1,426,603 games played, by October 2014 the network contained 420,000 terms and more than 12 million relations.

JDM is a type of game known as *open*; in this game, the players must type in their suggestions within a time limit from a target term and from instructions. This interaction modality strongly reduces the scenarios of usage, but is by far the richest of all the games offered. A JDM game, with a term and a type of relation, is played anonymously and asynchronously between two players. For reasons of safety and quality, we only store the common responses of both players. The relations thus acquired are weighted according to the number of pairs of players who have suggested them.

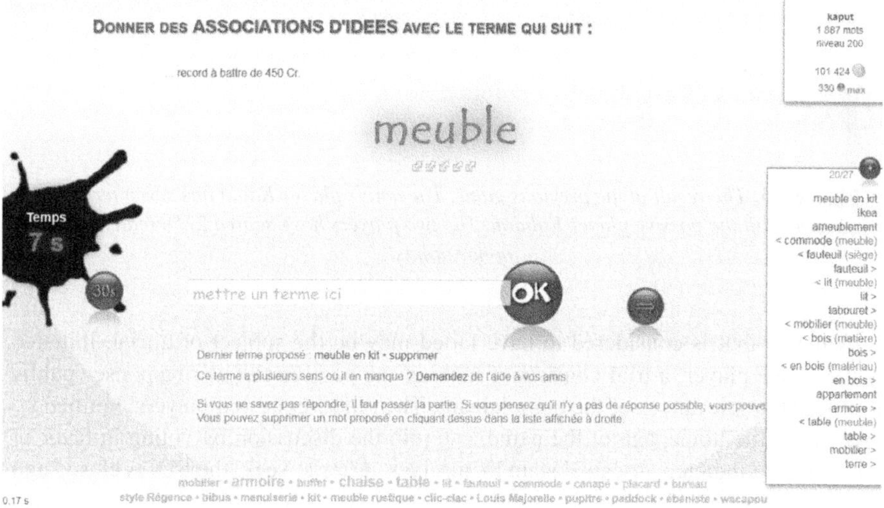

Figure 5.2. *A typical JeuxDeMots game. The player has to suggest terms that he/she associates with the target word "meuble" (furniture). His/her suggestions are shown on the right as he/she inserts them in*

Figure 5.2 shows a typical JEUXDEMOTS game in which it is asked to enumerate what may suggest the term *meuble* (furniture). The player has made a reasonable number of suggestions, and for some polysemous terms has selected the right sense. For example, among the possible uses for *commode* (chest of drawers), he/she has chosen *commode (meuble)* (chest of drawers (furniture)). The result of the game is presented in Figure 5.3. The list of suggested terms in common with the other player is displayed. Each of the two players (the active as well as the passive) collects the resulting points. The score being high, a number of actions are possible and constitute extra rewards for the active player.

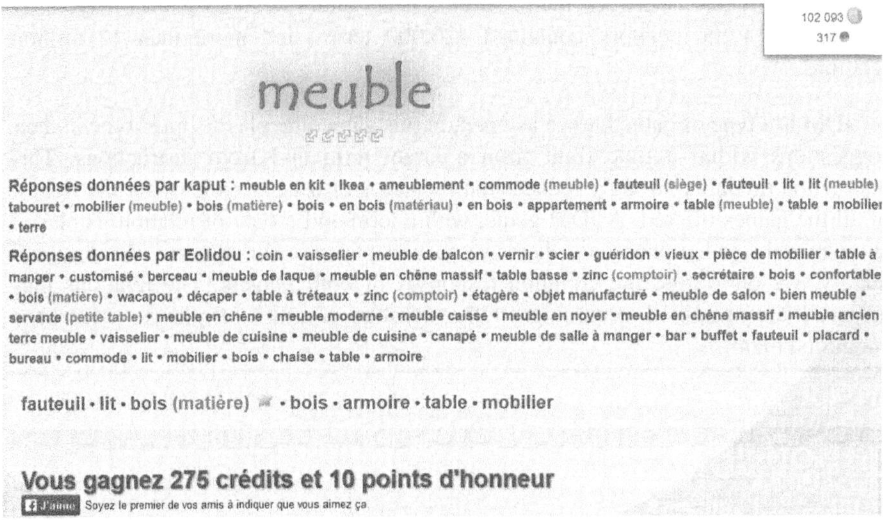

Figure 5.3. *The result of the previous game. The active player Kaput has seven terms in common with the passive player Eolidou. The two players have scored 275 credits and 10 honor points*

A match that is considered to have failed may be the subject of a trial. Initiated by the active player, a trial consists of making the passive player's responses public to expose the grievances of the complainant. The other registered players, notified by a post-it on the homepage of the game, can join the discussion by voting in favor of or against the indictee, or can abstain themselves. After a week, the trial is closed and the winner (the accuser or the accused) receives a symbolic compensation. The trials have a notorious educational effect, especially concerning difficult relations, mainly because of the space for discussion, where everyone can argue in favor or against and justify his/her vote.

Players also have the opportunity to offer gifts to other participants. A gift consists of the opportunity to play a game with a term-relation pair chosen by the player who is offering. The motivation here is essentially social: a player who offers interesting games is well received by the community. As a result, the development of the lexical network is accelerated and enhanced. The vast majority of players refrain from proposing term-relation pairs without interest. The analysis of the behavior of the players offering gifts shows that social pressure (through the forum or the trials) prevents sterile or unsportsman-like behavior.

Besides standard games, JEUXDEMOTS offers a multitude of game variants, which can be used to revive the interest of the player in order to break the monotony and to avoid boredom. Among other things, it is possible to:

– play with the themes of one's own choice, or select the ones of the other players;

– play on *easy* or *difficult* terms – the criterion of difficulty being automatically determined by the status of the lexical network;

– start *lotteries*, which enable the random pairing of two matches (a game of the player connected with that of another player); depending on the number of terms in common (*intersections*), several prizes of various kinds can be won;

– attempt to steal words from other players, challenge to duels, carry out missions, etc.

The purpose of all these game modes is to increase the information activity of the network. A large number of actions must be paid for (in the virtual currency of the game, the *credits* and the *blue pills*) and implicitly aim to create a situation of *famine* by encouraging the player to squander his/her loot. The player will not be able to escape famine unless he/she plays standard matches to rebuild his/her stock of *credits* and *blue pills*.

The games themselves are strictly cooperative because both players (active and passive) have exactly the same scores. However, many elements of the game encourage competition, in particular the presence of multiple rankings, the diversity of ranking criteria inciting players to play, both in the best manner, and as often as possible. Among ranking criteria, the number of possessed words (since the known ambition of the game is to *collect* words) may prove particularly motivating. The fierce competition between players is encouraged insofar as it is extremely rewarding for the network, since it implies that one must play better than the other. At the same time, the strategy of miserably playing to cause the downfall of the other is not viable since it leads to sacrifice one's own games and therefore also leads to fall in the rankings. Thus, malicious play can neither facilitate better rankings nor obtain any benefit, regardless of the circumstances.

Furthermore, JEUXDEMOTS is clearly a game of cooperation/competition. By playing well, not only is one's own overall score increased, but those of other players are favorably influenced. The diversity of classification criteria (points scored, number of games played, efficiency, etc.) strengthens activity while encouraging each player to play the best of their ability by favoring any strategy or game mode depending on the type of classification that motivates him/her the most.

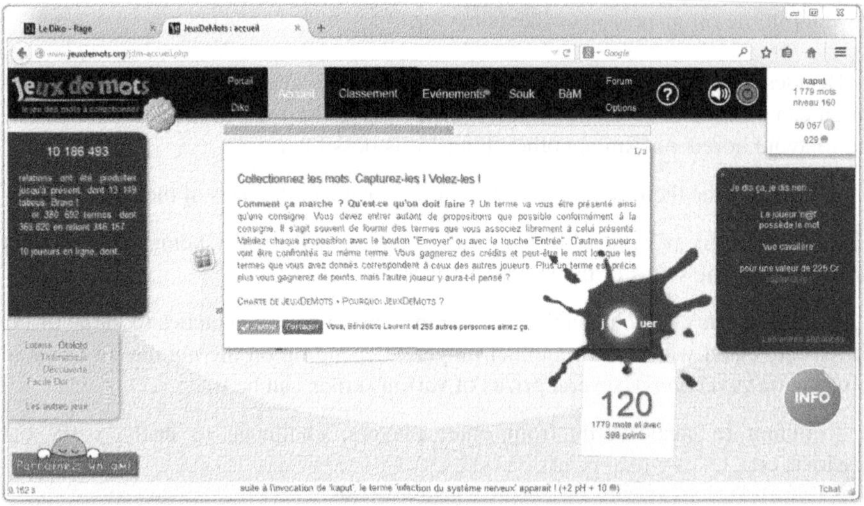

Figure 5.4. *Home page of the game JeuxDeMots*

The other (counter-)games are known as *closed* to the extent that one or more suggestions are displayed and must be selected or validated. In general, the games are shorter than those of JDM and can be played on smartphones or touch-screen tablets. This last point is far from being anecdotal insofar as many players indicated that they played these games during waiting periods (e.g. at public transport). The game TOTAKI, however, remains a partially open game because if the system does not find the correct answer, the player is invited to make a suggestion.

5.3. PTICLIC: an allocation game

The game PTICLIC [LAF 09], launched a year after JDM, aims at the consolidation of the network created by JDM: it makes it possible to strengthen or clarify existing relations. PTICLIC is a closed game of associations by drag-and-drop: a target word and a "cloud" of related terms appear, and the player must move the terms of the "cloud" one by one in the categories which seem relevant in comparison to the target word. For example (Figure 5.5), for the target word *restaurant*, should the term *commerce* (business) be regarded as a hypernym or a synonym? During a game, certain terms of the cloud are not associable to a relation and must be ignored. The cloud of terms is created in part on the basis of the JDM lexical network but also through a latent semantic analysis (LSA) analysis of a large corpus (journalistic and texts issued from youth literature).

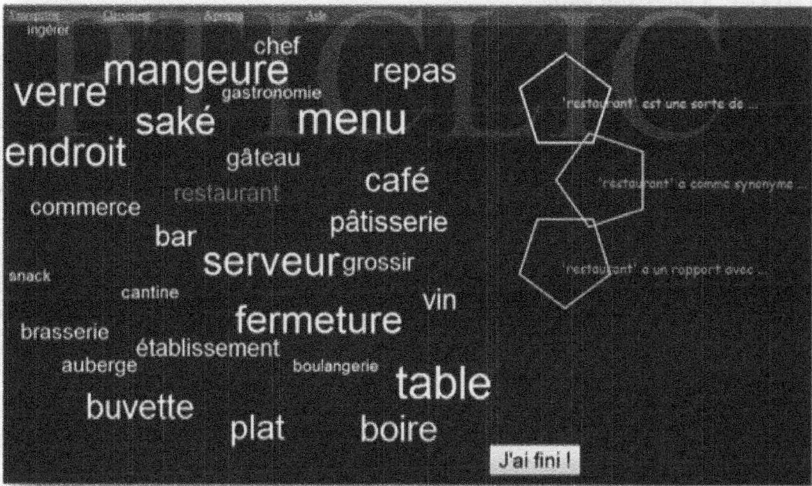

Figure 5.5. *PtiClic typical game. For the target word* restaurant, *the three target relations proposed are hypernymy, synonymy and association of ideas*

A PTICLIC game is not limited in its duration, but in practice it lasts less longer (between 30 and 45 s) than a JDM game, but longer than a clicking game. The modality of interaction makes the use of tablets and smartphones possible. On January 1 2013, about 60,000 PTICLIC games have been played since its launch.

The principle of PTICLIC is to display a target word and a series of words which are, within the lexical network, *associated ideas* of the target word. The player is then asked to define for these terms more precise relations with the target word, such as hypernymy (generic), hyponymy (specific), holonymy (whole), meronymy (part), etc. These are therefore ontological relations that are proposed as categories, strictly lexical relations (for example, *terms of the same family*) being much less suitable for this type of exercise.

The *associated ideas* relation remains proposed as the possible choice during games. As a result, certain terms of the "cloud" are extracted through methods based on LSA or in a more or less distant neighborhood of the target term in the network. However, it is not certain that these terms have a relation with the target term.

An extension of PTICLIC [ZAM 10] has allowed information to be gathered about the age of acquisition of vocabulary in young speakers. This type of approach can and should be applied in an educational context and proves useful in the analysis of the lexical difficulty of texts.

5.4. TOTAKI: a guessing game

TOTAKI, launched in early 2010, is not only considered a guessing game, but also as a tool to help with the problem of the Word on the Tip of the Tongue (WTT) [JOU 11]. Originally, TOTAKI has been thought and designed as a modality of qualitative evaluation of the JDM lexical network. However, the success of this game and the fact that many players are looking to trap TOTAKI have also led to game modalities which enrich the network.

The principle of the game consists of making the system guess a word by suggesting clues. After each clue, the system advances one possible answer. The clues can be typed; otherwise, by default they are considered as simply associated ideas of the word to be found. The type of a clue corresponds to one of the types of relations existing in the network (for example, a clue such as the generic relation: *is an animal* can be given, instead of giving *animal* only).

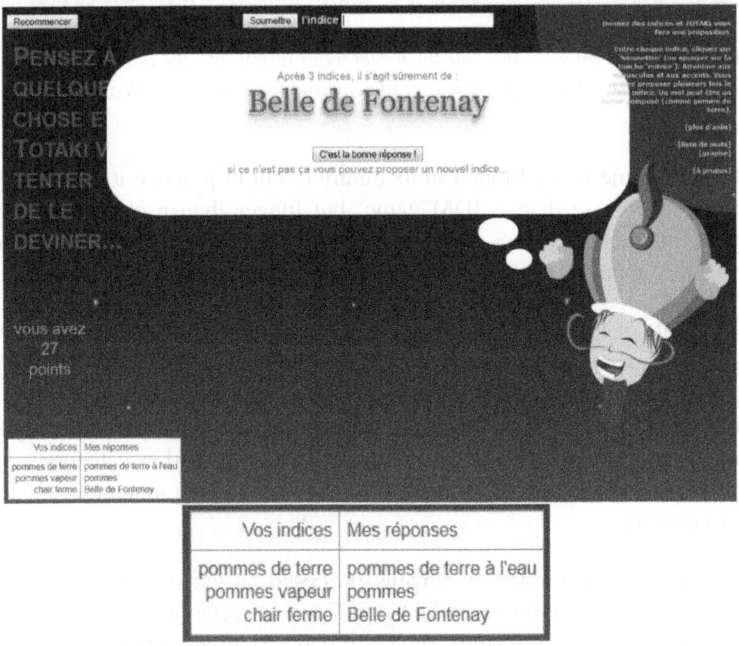

Figure 5.6. *Typical* TOTAKI *game. To make* TOTAKI *guess Belle de Fontenay, three hints were enough*

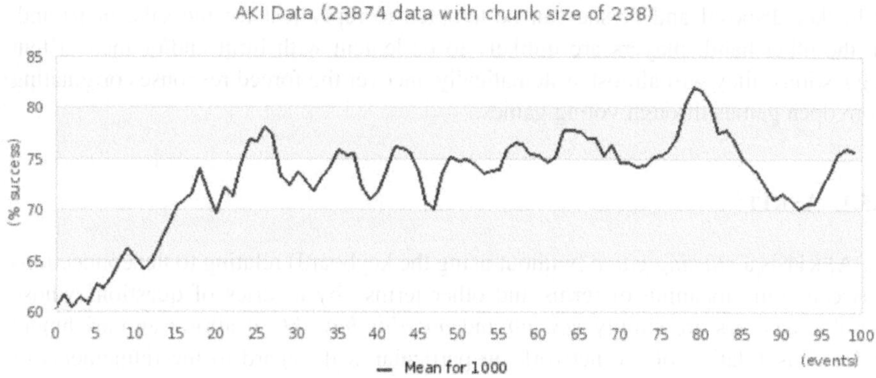

Figure 5.7. *Success rate of* TOTAKI. *Evolution of the success rate of* TOTAKI *to correctly guess the term targeted by the player*

The evaluation of the network through TOTAKI has been presented in [JOU 12]. The most notable result is that TOTAKI can guess the term targeted by the player 3 out of 4 cases. Under the same conditions, the human players succeed in 48% of the cases. In addition to an evaluation of the data obtained with the other games, TOTAKI also allows strengthening the network and detecting inconsistencies by a learning mechanism. Since its launch, more than 23,000 TOTAKI games have been played (the record for statistical purposes has been achieved starting from the 10,000th game).

5.5. Voting games

The principle of a voting game is to propose to the user to comment on a statement regarding the lexical data. It is a form of survey whose presentation assumes a ludic appearance. The vote is, in general, materialized by clicking on a suggestion among several (this will be referred to as a *clicking game* if there is interest in the modality). This type of game has the advantage of being very fast for the player and of providing a very fine granularity in terms of modification of the lexical resource. Following his/her selection, the player is informed of the manner in which he/she is ranked in comparision to other players. In other words, if he/she has replied as the majority of people or, on the contrary, if the choice he/she has made is statistically in a minority. This *response* from the game is the reward that awaits the player. The implicit goal of the game, implicit because not explicitly exposed to the players, is to obtain honest answers to the questions in order to ideally cause a consensus to emerge (if this is possible). Voting games are generally good counter-games to open games insofar as, on the one hand, they cannot give rise to *forced responses* (a forced response refers to the fact that the player can answer anything because he/she has an empty text field at his/her disposal and he/she can be tempted to reply just for the sake of it) and, on the other hand, players are unlikely to be lenient with limit and/or inconsistent suggestions: they will almost systematically uncover the forced responses originating from open games through voting games.

5.5.1. AskIt[2]

AskIt is a *clicking game* (without using the keyboard) relating to the connections between some meaning of terms and other terms: by a series of questions whose possible answers are simply *yes*, *no* and *possibly but uhh*, it allows an enrichment and a consolidation of the network, in particular with regard to the refinements of polysemic terms.

Figure 5.8. *Typical Askit game. On the left, the statement about which the player must give his/her opinion by responding "yes", "no" or "possibly but uhh", on the right, the result: The percentage of players of the same opinion being displayed, as well as the progress bar of the associated score*

AskIt is a closed and fast consensus game. The analysis of the motivations of the players shows that the excitement/frustration following a match comes from the hope that one has responded, or not, like the majority. The symbolic score in points is materialized by a progress bar. The implicit objective (i.e. to fill it) is a strong psychological incentive to play honestly. The trolling rate (players voluntarily responding with something erroneous) has been estimated to be less than 1% of the over 500,000 games played to this day. The question selection algorithm is twofold:

– selection by triangulation in the network;

The point here is to propagate relations by induction with expressions such as:

$$A \quad r_1 \quad B \qquad \& \qquad B \quad r_2 \quad C \qquad \Rightarrow? \qquad A \quad r_3 \quad C$$

2 AskIt is available at the address http://www.jeuxdemots.org/askit.php.

with specific selections of triplet (r1, r2 and r3). For example, if *canary is a bird* (a canary is a bird) and *bird agent sing* (a bird can sing), then do we have *canary agent sing* (can a canary sing)? We will observe that the answers to such type of questions are far from being systematically positive. As a result, a counterexample would be: *stork is a bird* and *bird agent sing* =>* *stork agent sing*[3] (storks are included in the list of very rare birds that do not sing). Moreover, we have particularly focused on the refinement relation, thus making it possible to cause the relations to switch from a polysemic term toward one or more of its usages:

frigate → *refinement*→ frigate (bird)
& frigate → *partof*→ hull (*frigate* has *hull* as a part)
⇒* frigate (bird)[3] → *partof*→ hull (It is WRONG to say that *frigate (bird)* has *hull* as a part)
frigate → *refinement*→ frigate (ship)
& frigate → *partof*→ hull (*frigate* as *hull* as a part)
⇒ frigate (ship) → *partof*→ hull (it is right to say that *frigate (ship)* has *hull* as a part)

– selection of weak relations.

The second method consists of asking for confirmation (or invalidation) for occurrences of relations with a low weight, and therefore subject to caution.

Triangulation is a simplistic approach that gives about 60% of correct results. This means that approximately 40% of the suggestions made to the players are wrong. We will notice that this is desirable and voluntary. In fact, a *too good* algorithm, which would generate suggestions giving rise to a majority of positive responses, would be annoying for the player. The significant proportion of wrong but interesting suggestions maintains the attention of the player and his/her interest in the game.

Moreover, we try to trap the player by reversing suggestions that proved to be correct. This makes it possible to bring a little more variety in the answers and above all to detect the players who answer randomly or systematically.

A *no* response slightly decreases the weight of the relation at stake. The weight of some relations can thus become negative (they are then known as *inhibitors*). For example, *ostrich agent fly* (an ostrich can fly) has a negative weight of −45 in the database (at the time of the writing of this chapter). These inhibitory relations are particularly interesting in semantic analysis because by playing a prohibition role they stop the propagation of the signal on a network of occurrences of terms. About 10,000 inhibitory relationships have been obtained by ASKIT in 12 months.

ASKIT has a number of notable characteristics, namely:

– it is an addictive game: questions automatically follow each other without any intervention from the player, who finally has always a question in sight that he/she can

3 * designates a false assertion

answer with a click. The game modality is so simple, fast and fluid that even the player who would have decided to stop can be easily driven to play a few additional games, and this all the more as he/she sees his/her score progressing between two questions;

– it can be fun, even catchy, especially with polysemic terms (e.g. can a *dog (tarot) bark*?);

– the games are very short: they last only for a few seconds. However, the game sequence can be long (more than half an hour) as the players themselves admit;

– the answers are spontaneous: is it better if players have no particular expertise in linguistics? This generates data from common general knowledge and not expert knowledge. It should be noted, however, that if the question is too complex or ambiguous, this can result in a large number of non-responses;

– the result (how is my opinion compared to that of other people, and am I in agreement, or on the contrary, in disagreement with the majority of the other players?) is immediately given: the player feels valued by a response conforming to the majority response ("I have correctly answered"); in the event of more complex questions where no clear majority emerges, he/she can tell himself/herself "I am smarter than them", if he/she has hesitated and has the feeling of having provided a clever or thoughtful response.

5.5.2. LIKEIT

LIKEIT[4] tackles simplified *feeling*-based relations, with relatively simple questions asked to the player, based on the model "*do you like the idea of*" followed by a term. The only possible responses are *yes*, *no* and *neutral*. LIKEIT thus enriches the lexical network in terms of *polarity* terms. The preliminary use of this data, within the framework of a lexical disambiguation, seems to show that polarity taken in isolation allows selecting the proper use of a term in context in 50% of cases. This type of polarity data can also be employed in opinion analysis.

LIKEIT is a consensus game whose properties are developed below. With regard to its interest and qualities, a certain number of notable features can be observed:

– similar to ASKIT, LIKEIT is an addictive game, with short matches and immediate restart;

– the answers are very simple: positive/negative/neutral– they are similar to those required in a survey, however, the diversity of the proposed vocabulary and the topics it tackles is such that, according to the feedback, the players do not have the impression of responding to an opinion survey;

– some terms raise very mixed feelings (for example, the term *operating bloc* is positive in absolute terms but negative when it personally concerns someone); the

4 LIKEIT is available at http://www.jeuxdemots.org/likeit.php.

opinion of the same player can change over time, depending on the circumstances (e.g. *high school diploma* arouses a negative feeling if one is a high school student, but much more positive when it has been obtained).

Figure 5.9. *Typical LikeIt game. Example of two consecutive LikeIt screens. Following the response in the left screen, the player sees immediately at the top of the next screen (right image), the percentage of players who share his/her opinion: direct feedback is thus given, while the game is immediately rerun with a new question*

The selection of terms to propose to the player is carried out pseudo-randomly in the lexical network of JeuxDeMots according to a propagation approach. A term already partially positively or negatively polarized (but not only neutral) is selected at random and one of the terms to which it is connected is also selected. Thus, this very simple algorithm (known as *random walk*) performs a propagation within the graph between potentially interesting terms for the acquisition of polarity values. A strictly random selection within the lexical network would give a too high proportion of neutral terms which would diminish the interest of the game. In 3 months, more than 24,000 terms have been *polarized* (i.e. owning polarity information) for a total exceeding 150,000 votes. In three years, almost 100,000 terms have been *polarized* for a total of 5,780,790 votes[5].

The game LikeIt presents a possible bias: for a polysemic term, it is possible that the response of the player undergoes some *contamination* by a strongly polarized anecdotal sense. For example, *cow* whose main meaning, the animal, is overall neutral (or even slightly positive) may be contaminated by a sense of *cow (vicious)*, which has a strong negative polarity. Thus, the players, being *de facto* in a polarity context through the game, will vote thinking about the most polarized meaning, and therefore choose a negative polarity for the term *cow*.

5 http://www.jeuxdemots.org/likeit.php?action=list.

5.5.3. SEXIT

The game SEXIT [LAF 14b] is very similar to ASKIT and aims at annotating terms based on their actual or perceived relation to the concept of sex. This type of information finds usages in document filtering, especially those of pornographic nature. The instructions are deliberately ambiguous, in that the notion of sex can both refer to biology and reproduction, as well as to behaviors or sexual practices, or to content deemed licentious. It is therefore certain that regardless of the quality of the data collected, postprocessing is necessary if the different uses are to be distinguished.

The first version of the game proposed to the player, after about 20 games, a humorous profiling such as *obsessed maniac/wide eyed*. This profiling, which had no psychological pretensions, was simply intended to amplify the ludic aspect of the game. The calculation of the profile was relatively trivial: the player would be qualified as *maniac* (respectively, *wide-eyed*) if he/she had a majority of responses of the *sex*-type (respectively, *no sex*) for terms that other players would mostly find *no sex* (respectively, *sex*).

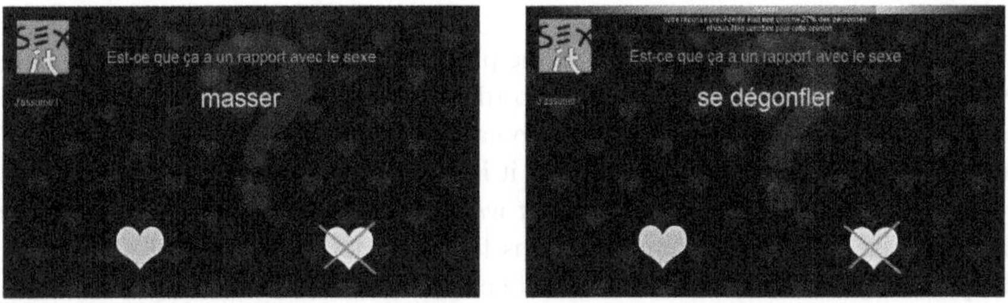

Figure 5.10. *Typical* SEXIT *game. The player must choose between two possibilities: has the word any relation with sex (non-barred heart) or not (barred heart). By clicking, one is directly brought to another game, while the result of the previous choice (percentage of players of the same opinion) is displayed at the top of the screen*

At the end of approximately 5,000 matches on about 500 different terms, it was found that the data were strongly biased. More specifically, terms with *a priori* no relation with the concept of sex had, on average, one-third of the votes for a sexual connotation (*sex*). Symmetrically, clearly sexual terms had approximately one third of votes *no sex*. This unexpected percentage of votes was far too high to be attributed to a natural variation due to errors or to a few players practicing trolling.

The behavioral analysis of players-testers has showed that about one-third of them tended to select their responses based on what they wished to obtain as qualifier, namely *maniac* or *wide-eyed*. Therefore, the choice was no longer based on a personal appreciation but rather on an objective to achieve *a priori*, namely to be cataloged either as a *sex maniac* or as *wide-eyed*, etc. Completing this goal was immediate by systematically clicking on one of the two choices, without even having to think or even read the suggested term.

Updating SEXIT by removing the mentions of *wide-eyed* and *maniac* and by resetting the data has immediately solved the problem. For the data obtained with the new version, the noise/variability does not exceed 5%.

Here, it consists of a typical example where the game loop, poorly analyzed, originally altered the quality of data produced by falsifying them. The objective set by the game assumes the creation of quality resources by the player, that is to say, it faithfully reflects his/her language practice. Any other objective that the player would be tempted to achieve should be neutralized to the extent that it is likely to bias the data.

Another possible bias with SEXIT would come from the fact that men and women do not understand the instructions in a similar way. Women would tend to interpret the instructions so as to answer "yes" only when the suggested term invariably causes thinking about sex regardless of the context. Men, for their part, would tend to answer "yes" if the term may eventually have some link with sex. It can be estimated that these biases compensate each other.

5.6. Multi-selection games

Multiple selection games give the opportunity to respond to instructions by selecting several suggestions, and even for some games to insert one's own responses through a text field. Therefore, it concerns an intermediary game modality between JeuxDeMots and voting games.

COLORIT – The game COLORIT[6] gives the opportunity to the player to associate one or several colors with the displayed word [LAF 14a]. The player may also pass or associate a *non-color* information (notably for abstract terms). The player can select different categories of terms: *simple/hard*, *animals* and/or *plants*.

The game prompts the player to be effective and exhaustive, that is, to suggest several colors, although this is less immediate than to simply click on one of the 15 colors of the palette. While playing, one accumulates points and accesses levels. The higher the level, the more characters can be inserted into the text field. Thus, the frustration of being limited in one's responses because of a low level (the player can enter only one color name or two) prompts the player to continue to play in order to be able to suggest more answers. The possibility to deviate from the proposed palette allows to collect more accurate color names than the 15 possible default choices.

6 COLORIT is available at http://www.jeuxdemots.org/colorit.php.

The notification of the result displays as a percentage all the colors that have been proposed by all the players for the term. In addition to the score in points (all the more higher that the given color(s) come(s) closer to those selected by others), these data constitute another form of a reward for the player: not only does he/she find out the colors that others associate with the term, and to which he/she eventually did not think, but he/she also visualizes how his/her own answers compare.

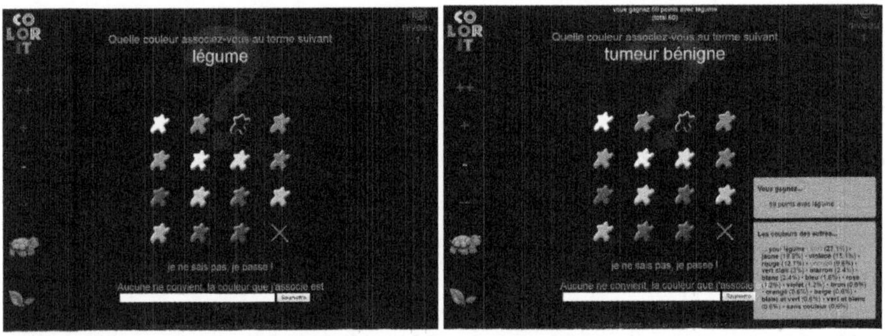

Figure 5.11. *Typical* COLORIT *game. For the word* vegetable*, the player must either click on the color that he/she most spontaneously associates with this term, or suggest several of them and/or some others through the text field. His/her response causes a new term to be displayed with the score in points at the top of the screen and on the right the answers of the other players for the previous term, with their percentages*

About 800 terms related to colors or to aspects have been collected through COLORIT. Around 18,000 words have been associated with these color terms, resulting in about 32,000 relations through 430,000 votes.

EMOT – The game EMOT[7] is based on the same principle as COLORIT and shares the same operating mode and interface, but this time the player is invited to state his/her opinion on the emotions/feelings that he/she associates with the target term (Figure 5.12). Again, the interest of the game lies in the comparison of the player's own responses with those of the other players. The possibility to choose between easy terms (usual) and difficult terms (less frequent) adds some spice to the game.

The selection of the suggested terms is automatically carried out according to a *propagation*-based algorithm, whose simplified principle is the following: a term already associated with an *emotion* or a *feeling* is randomly selected in the lexical network. Then, the algorithm proposes in an equiprobable manner either this term or a term that is linked to it in the network. Thus, as the number of games (of matches?) increases, the terms are progressively associated with values of *emotion/feeling*. This approach, although it can't avoid it, helps to minimize the risk of proposing too often

7 EMOT is available at the address http://www.jeuxdemots.org/emot.php.

terms with little or no relevance for the association with a *feeling* or an *emotion*, and thus terms of little interest in the context of this game.

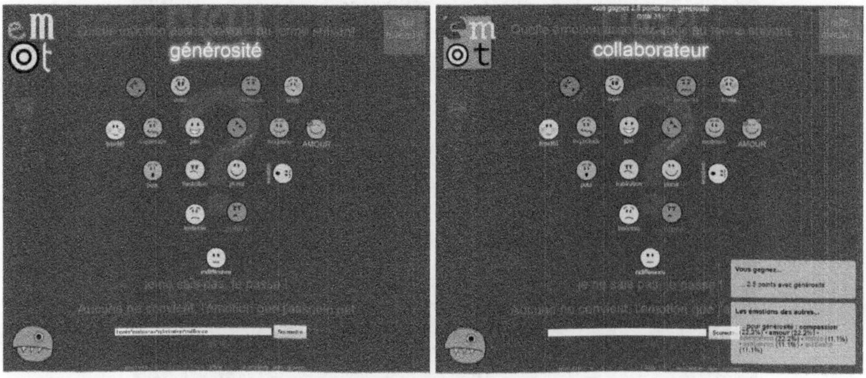

Figure 5.12. *Regular* EMOT *game. The player must associate one or more emotions/feelings with the term generosité, either by clicking on a smiley or by entering words in the text area. As for* COLORIT, *the answer causes the suggestion of a new term, with the score in points being displayed together with the responses of the other players and their proportions*

About 1,400 terms designating an *emotion* or a *feeling* have been collected with this game. About 7,000 words have been associated with these *emotion/feeling* terms, which has created around 20,000 relations from 1,500,000 votes.

POLITIT – The game POLITIT[8] is relatively similar to the previous two games. However, in this game, the players are invited to give their opinion on the *political color* that they associate with the target term by selecting only one from a list of closed political choices (far-left, ecologists, moderate left, center right, conventional right and far-right). The choice triggers the suggestion of a new term and the displaying of three types of information about the previous vote: the percentage of players who made the same choice, the distribution of the votes of the other players among the six possibilities and the broader right/left distribution.

The terms choice algorithm is similar to that used in the game EMOT, except for the selection criterion which has to be connected to a value designating a *political tendency* (instead of *emotion* or *feeling*).

TIERXICAL, a sorting game – In TIERXICAL[9], the player is prompted to rank by order of relevance the three terms which, in his/her opinion, best suit the issue. The

8 POLITIT is available at http://www.jeuxdemots.org/politit.php.

9 TIERXICAL is available at http://www.jeuxdemots.org/tierce/index.php.

design and the interface of the game clearly refers to a *lexicalized Pari Mutuel*, where horses would be terms associated by a relation with a target word.

Figure 5.13. *Regular* POLITIT *game. The player must associate the term race with a political tendency. Only one choice is possible. The bottom image shows the percentage of players who have made the same choice, and the distribution of the responses of the other players between the six possibilities, and between right and left (here, a particular case showing a unanimous choice)*

The scored points depend on the correspondence between the choice of the player and the status of the lexical network. The interest of this game is to adjust the weights of the relations already present in the network on the basis of the opinion of the players. A player who achieves the *trifecta* has not only chosen the three most associated terms (therefore for which the weights of the relations with the target are the highest), but he/she has also correctly classified them, that is, in decreasing order of weight. Thus, the *straight trifecta* will have the effect of confirming the status of the network, increasing the respective weights of the three relations. But in the case of a *boxed trifecta* (the player has chosen the three most associated terms, but not in the correct order), or a choice not corresponding to a *trifecta*, the player scores less points, but his/her answers have exactly the same impact on the status of the network: the first choice increases the weight of the relation by five points, the second choice by three points and the third choice by one point. The calculation and the allocation of points is only a game effect to motivate the players to make the choices that seem the most honest to them. The game is exciting because it is akin to betting on what the other players have placed on average.

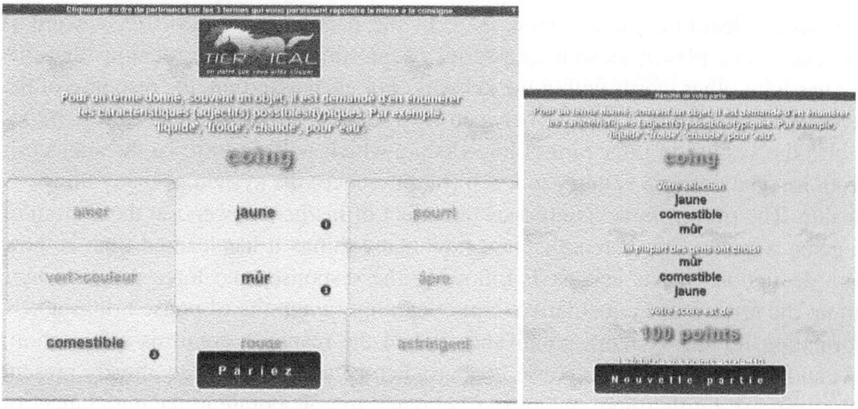

Figure 5.14. *Typical* TIERXICAL *game. Among nine possible characteristics, the player is prompted to select in the order the three terms that he/she considers most relevant of characteristics for the term quince. On the right, it can be seen that the result of this game is a boxed trifecta*

The player also has the option to issue *negative* choices by clicking again on a chosen term, if he/she considers that some of the proposals are not appropriate. In this manner, the player enters an advanced game mode, where the objective is to select the three least suitable proposals. This possibility allows either to introduce or to strengthen negative weights for certain relations (indicating, for example, that an *ostrich* does not *fly*, or that *ice* is not *hot*).

Since the launch of the game in mid-2012, around 35,000 parties have been played and have, therefore, actively participated in the adjustment of the weights of the relations in the lexical network. None of the games is stored; the comparison between the responses of the player and the status of the network is done synchronously.

5.7. From games to contributory systems

The experience of the project JEUXDEMOTS has shown that some very regular players, having developed a good knowledge of the workings of JEUXDEMOTS, manifested interest and curiosity about the project itself, and wished to contribute more directly to the creation of the lexical network. These players do not stop playing so far, but their profound interest in language and vocabulary drives them to try to understand how the game builds the network and to dedicate themselves in another manner. The pride in participating in a collective project, in the constitution of a lexical resource freely available, becomes for these players a motivation element that takes precedence over the attractiveness of the game as simple entertainment.

ASKU, a learning pseudo-game – In the pseudo-game ASKU[10], a word is presented to the player, as well as a series of questions about the relations that this term has with other words within the network. For each assertion, the response of the player, who will have to choose between *true/possible/little relevance/uhh no/false,* enables the system to infer further knowledge, which will, in turn, be the subject of questions. The purpose is to try to teach (by inference) the system as many things as possible. The player immediately sees the effect of his/her answers: at the bottom of the game window, the system displays how many things it has learned (that is, how many deductions it has achieved) following the response, and three progress bars sum up the responses in qualitative terms: learning (when the response brings a new information for the system), connivance (when the response confirms information) and annoyance (when the response is contradictory with knowledge already present in the system). Furthermore, the player is invited to collaborate to the enrichment of the network in an open manner by entering information about various relations of the target term through a free text field.

Figure 5.15. ASKU *screen. The player/contributor is invited to give an opinion about a number of suggestions relating to the term felicity in the sense of happiness. For each suggestion, a gradation of possible responses is available, ranging from the unreserved consent true to total rejection false*

10 ASKU PSEUDO-GAME is available at http://www.jeuxdemots.org/askyou.php.

DIKO, a contributory environment – DIKO[11] is a web interface for consultation and contribution associated with the lexical network of JEUXDEMOTS. The user can consult terms associated through various relations with the term of their choice by means of a form. For every relation, the associated terms are listed in decreasing order of weight.

The interface of DIKO makes it possible to switch into a *contributive* mode. Each entry in DIKO corresponds to a term to which the user can associate other words, for approximately 80 relations. Each term then associated by a relation constitutes a *contribution*. The contributions are visible to other contributors, who can vote in favor or against. Administrators will be able to subsequently validate or invalidate the pending contributions. The guiding principle of the design of DIKO is to place users/contributors in a situation of negotiation with other users. For each contribution, a *discussion thread* can be opened either by the author of the contribution or by another participant to express a doubt, disagreement, ask a question, justify a negative vote, etc. This approach by negotiation is opposite to that of JDM, the main game, where players express their choice in an independent manner and where the input of new data into the network is the result of a confrontation carried out by the system (it should be noted that to insert a relation in the network through JDM, it is necessary that the corresponding term has been proposed by the two players whose responses are compared during a game). In the contributory approach of DIKO, the insertion of a new relation is known as *negotiated*: the contributors are the ones who debate and vote, and a human administrator moderates.

The interface of DIKO is poorly gamified; the only concession to the ludic dimension is the display of a *Hall of Fame* ranking the contributors on the basis of the significance of their participation. A player who achieves a validated contribution scores three points while voting for such a contribution gives one point. The more a contribution has positive votes, the more likely it is to be validated and, therefore, to bring points to its author. This mechanism drives the contributors to be both verbose and cooperative. As a result, the more a contribution will appear relevant, the more it will collect positive opinions (*in favor* votes). Playing solo without voting for the contributions of others is not productive, nor is only being a follower. It should be noted that for contributors, the existence of the *Hall of Fame* as *classification* has no essential relevance, the primary motivation remaining the interest in the project and in the lexical thing.

The most active players/contributors have thus won hundreds of thousands of points. An estimate would be that only about 400,000 relations would have been introduced through the contributory mode of DIKO. It should be noted that most of the time it concerns contributions particularly interesting for the enrichment of the

11 DIKO is available at http://www.jeuxdemots.org/diko.php.

network. Moreover, these contributions would have been difficult to obtain in a non-negotiated manner with games.

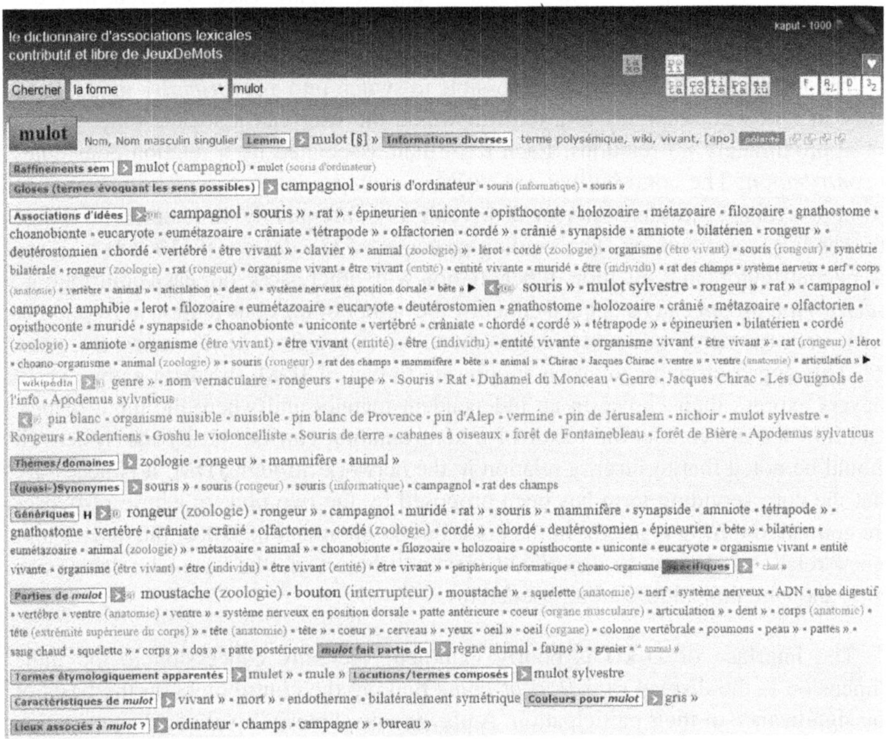

Figure 5.16. *Typical* DIKO *screen. The user has displayed the entry* mulot *(mouse), and the terms related to* mulot *by various relations can be seen. The interface is configurable and the level of detail (number of words to display, accuracy of the information, etc.) can be adjusted by the user*

5.8. Data collected and properties of the games presented

As a conclusion to this chapter, we mention the collected data, as well as a few properties of the games we have presented. By analyzing the data from the network JEUXDEMOTS, it is easy to unveil the difficulties that players face with certain types of lexical-semantic relations. It should be highlighted that they naturally avoid playing the relations they find too difficult, or uninteresting. The same behavior is noted when it comes to contributing to the resources with the DIKO interface.

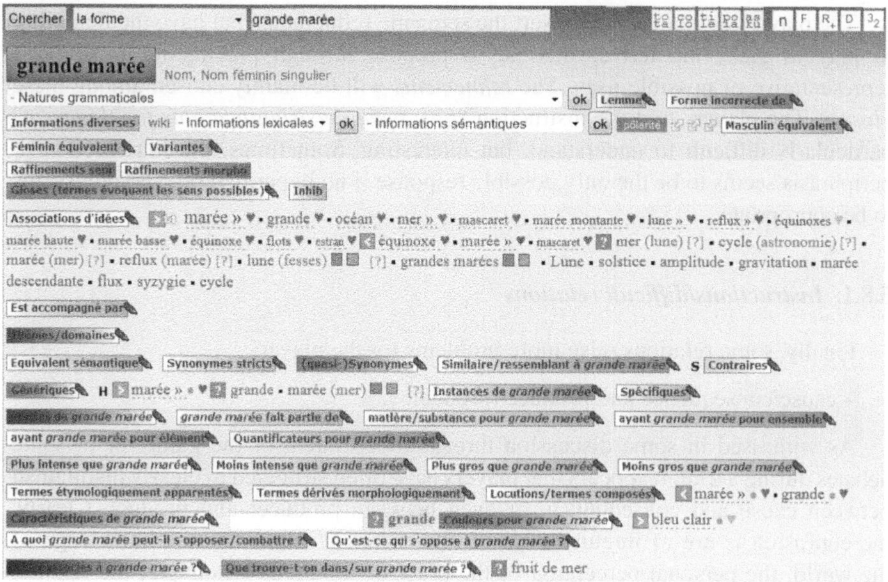

Figure 5.17. DIKO *screen in contribution mode. The user has displayed the entry* grande marée *(big tide), and the terms proposed but not yet validated can be seen (these are the words followed by little squares used to vote for or against these word as relevant or irrelevant proposals). The small hearts correspond to the polarity. The interface is configurable and the colors of the labels of the relations have been adjusted by the user*

Since the launch of the project JEUXDEMOTS, the lexical network has acquired more than 12 millions of relations. The initial database of 150,000 terms currently exceeds 420,000 (on October 1 2014). There are approximately 80 lexical-semantic relations in JDM, and most of them could be included in JDM games, or in questions in the counter-games.

Some of the relations of the lexical network are not playable, notably:

– 520,148 morphosyntactic categories: playing on the parts of speech would be of no interest;

– 78,312 lemmas, obtained from automatic or contributory manners (off games);

This information cannot be acquired using games because they are not sufficiently variable to present a ludic interest. It is therefore necessary to insert them manually or by any other (semi-)automatic ways;

– 22,842 semantic refinements originating from the calculation of clicks [LAF 10] and/or from the contributions of players.

The players do not directly insert the semantic refinements but have the flexibility to play to *gloss* the terms, that is, to propose through playing terms that are representative of possible uses. The *refinements* will be mainly chosen among these *glosses*. The players find the instruction "*give meanings/significations for this term*" particularly difficult to understand, but interesting. Sometimes, the introduction of periphrasis seems to be the only possible response if no hypernym or synonym seems to be appropriate.

5.8.1. *Instructions/difficult relations*

Finally, some relations raise more problems for the players:

– cause/consequence and producer/product:

As witnessed in some discussion threads independent of the game, or in some debates during a trial, it appears that players have often struggled to clearly distinguish between cause and consequence, or even between producer and product. Clearly, the confusion is not of linguistic nature, but rather in relation to the knowledge of the world, the personal perception of the order of things. For example, the relation *cause/consequence* (or *consequence/cause*) seems difficult to establish between the terms *fever* and *patient*.

By crossing, it is sometimes difficult to choose between *producer* and *cause* on the one hand, and *product* and *consequence*, on the other hand;

– the difference between "instance" and "hyponym":

For many players, hyponyms seem to include instances (but not the other way around). For example, *Jolly Jumper* is seen as an instance of *horse* as well as a hyponym. However, *thoroughbred horse* is (rightly) regarded only as a hyponym;

– the difference between the agentive role and the stages of construction:

The agentive role refers to the way in which an object is created/built. Accordingly, construction verbs of the entity/artifact are required; for example, *construct* for *house*, or *write/edit* for *book* [PUS 93]. The agentive implication concerns the verb(s) that is(are) involved in the construction process of the object (for example, for a *cake*: *mix*, *place into an oven*, *bake*, *remove from tin*, etc.). The distinction between the semantics of each of the two relations is not immediately intuitive for the players which consequently cause a significant number of incorrect games;

– the distinction between "magn" (more intense) and "good" (ameliorative):

The axis of intensity relative to a term is often perfectly subjective for the players, who often, next to the responses that perfectly respect the statement *is more intense than* (of which a canonical example is *strong fever* or *raging fever* as magn of *fever*), indicate terms that are of the ameliorative kind (e.g. *excellent meal* is more of an ameliorative than a magn of *meal*);

– feelings/emotions:

The concept of feeling/emotion sometimes seems very extensible, and it happens that some players assign the value of feeling or emotion to terms that are related to the association of ideas. For example, we will find *wealth* as feeling/emotion associated with *gold (money)*. The term *wealth* then carries the meaning of *feeling of wealth*;

– terms of the same lexical family:

Initially, the goal was to obtain words morphologically derived from the target term using the games on this relation. The instruction "*give words of the same family for the term that follows*" has been interpreted by a majority of players both in the sense of *etymological* family and *morphological* family; people do not make a clear distinction between the two concepts. It should be noted that a few disagreements on this issue have given rise to lively debates in the context of trials or discussion threads, and have had the merit to confer some expertise to a few passionate players. Sometimes surprising, the results demonstrate not only this amalgam of etymology and morphology, but also the confusion related to homophony. Thus, for the substantive *nuit* (night), we obtain *nocturne noctambule*, *noctambulisme*, *noctuelle* and *noctule* (nocturne, noctambule, noctambulism, noctuid and noctule), which are etymologically related; *nuitée, nuitamment* and *minuit* (night, nightly and midnight), which are morphologically derived, but also intruders such as *nuisible* and *nuire* (nocive and harm), which share a homophone part with night; and, finally, *chemise de nuit* and *bonnet de nuit* (night gown and night cap), which are only composed words including the target word. Sampling estimates about 5% incorrect associations regarding this relation (errors being sometimes widespread among speakers). To summarise, etymologically related words and morphologically derived words are validated as *terms of the same family*, the distinction between the two types occurring downstream in an automated way;

– problem with typicality:

For some relations (such as *agent*, *patient* and *characteristic*), we were expecting *a priori* to obtain *typical association*. For example, *to roar* is a *typical* action for a *lion*, while *running* is *possible*, but not specifically *typical*. Similarly, *tiger* is typically *ferocious*, but can also be *old*. By examining the data produced by the players, it turns out that all possibilities (typical or atypical) tend to appear in the network, but with much smaller weights than those of typical associations. This is a good news because where we were only expecting to collect *typical* information, we also collect *possible* information, and therefore additional data, with weights allowing for a clear distinction.

5.8.2. *Forcing, players typology and error rate*

Should a player who provides an incorrect answer be penalized? We do not believe that it should be the case. First, it would require a precise definition of what constitutes an erroneous relation, accompanied by an effective and unambiguous detection mode. Moreover, the risk of being penalized can be easily inhibiting, or even downright dissuasive in cases of doubt, while repeated penalties can result in a lack of motivation and a disaffection for the game, both effects being counterproductive.

One of the major biases of JDM is an overall tendency of the players to do some forcing, that is to say, to respond at all costs – as though it were absolutely necessary to find and give answers. Although the player has the possibility either to indicate that there is no compatible answer with the word and the instruction, or to pass if he/she is not inspired, it may happen that he/she still prefers to indicate what seems the closest. Therefore, this is the reason why for specifics of *week*, the days of the week have been proposed: *Monday, Tuesday*, etc. The relations relative to this phenomenon of *forcing* have been assessed (by sampling) to approximately 0.1% of the generated relations. It is worth noting that the presence of counter-games largely compensates this bias specific to open games.

That being said, we have noticed that spaces of exchanges between players (trials, forums and, to a lesser extent, the instant messaging window) are privileged places to clarify the semantics of the relations when it is not well understood by players. As a result, an assiduous player will see his/her expertise increase significantly over time.

We have more accurately assessed the overall performances of the players. In JDM, each game in stock is "signed" (we know which player is its creator) and the average number of common responses between the two players can be incrementally calculated (it will be referred to as *performance*). Moreover, with the agreement of volunteer players, we have a partial idea of their profile (age, sex, education level, areas of expertise, etc.). This has allowed us to highlight the following facts.

The average efficiency of the players is 2.7 (that is, they average 2.7 terms in common). The best players have an efficiency around 5 or even 6. Players whose professional activity falls within the field of linguistics in the broad sense (phonetics, linguistics, computer science-linguistics, document management, etc.) have an efficiency of 1.8 with all the other players and 2.2 between them. The analysis of their games shows that "linguists" make fewer suggestions than the majority of the other players, and their suggestions are mainly among the most activated terms of the network. Thus, most of the time, it concerns correct but hardly original proposals and not feeding the long tail[12]. The players with a professional activity related to language (linguists, French as foreign language, NLP, etc.) represent approximately 8% of the players and 6% of the activity.

A comprehensive statistical evaluation was used to estimate the number of relations that are obviously erroneous (this was found to be 0.5%), and those that are

12 http://en.wikipedia.org/wiki/Longtail.

questionable (this was found to be less than 2%). Approximately 0.4% of erroneous relations are automatically detectable. This is the case, for example, of double hypernyms (A hypernym of B and B hypernym of A) which may raise problems, unless words are polysemous. False negatives are estimated at around 40%. The remaining 0.1% should be discovered on a case-by-case basis but concerns low-weighted relations and tends to naturally decrease with counter-games. The TOTAKI experience has shown that from a game of riddles, and strictly from the JEUXDEMOTS network, the system found the right answer in 75% of cases (compared to 48% for human interlocutors).

In conclusion, it seems important to note two things. All the games of the JDM portal aim to build a resource by consensus. This means that unlike conventional question games (such as a quiz game), there are no good (correct) or bad (false) answers. What defines a valid response is the number of players who consider it as such, since they have provided it. Thus, the point score is only dependent on the fit between a given response and all the answers of the other players to the same question.

Finally, the data of the JDM lexical network, resulting in large part from the activity of the players, are widely accessible and royalty-free. They can generally be recovered on the basis of a monthly export or dynamically term-by-term.

Conclusion

What are the qualities and advantages of a good GWAP?

In light of our experience with the JEUXDEMOTS project, as well as of the analysis of a number of other games, we will attempt to define the necessary characteristics for a successful GWAP. These can be found in the answers to a few simple questions: what do we want to ask players, what can we ask them and in what form? Almost all of the elements of the analysis grid that follows, built from the games of the JDM project, can be transferred to non-lexical games. Very few studies are available about GWAPs and most are related to a very general context [AHN 08a] or rather constitute a description of what is available [THA 11].

Players

In order to attract players, and then encourage them to participate, a game must have a certain number of qualities (obviously non-exhaustive list):

– fun:

The game must present a ludic interest (is it not obvious?) at the interface level to attract gamers, but even more at the content level in order to keep them. The content must be able to renew itself with a small dose of repetition offering the player the possibility of amending himself/herself in the event of failure. In the case of the construction of lexical and terminological resources, the large size of the vocabulary offers almost unlimited opportunities for renewal. This is also true in other areas, as we have seen in a number of GWAPs: for example, the number of characters that one can attempt to make AKINATOR guess is potentially unlimited;

– easy to understand:

Many potential players do not read the instructions; it is therefore necessary that the game be intuitive and that learning be as short as possible. In the context of a lexical game, a large amount of information and/or instructions is directly understandable by the speaker and does not normally require a developed linguistic background. Many other GWAPs do not require any training and are immediately accessible: for example, the player has only to know Tetris to be able to play PHYLO; no learning is required to find birds and their eggs in the NIGHTJAR project;

– immediate restart with *reclick*:

In addition to the intrinsic interest of the game, players play all the longer as the game can be immediately accessed and instantaneously restarted. It results in a quantitatively enhanced acquisition of resources. Minimizing the delay and the number of clicks needed to launch a new game is a strategy to favor as much as possible;

– evaluation of players:

A system of points and ranking between the players generally makes the game more attractive: establishing and stimulating the competition between gamers drives some of them to spend more time playing, thus acquiring practice and experience. This makes them not only more productive, but overall increases the quality of the contributions. It is very rare for a player to be put off by the presence of a ranking. In the worst case, he/she can choose to ignore it. In addition, in games involving a confrontation with the system, players can be put into a situation where they feel *smarter than the machine*; for example, in TOTAKI [JOU 11] of the JDM project, a large number of games are intended to trick the system, or to teach it something. Players then have the (legitimate) feeling of being useful in the process of acquiring knowledge. The designers often play on the heartstrings of *citizen science*, for example in WORM WATCH LAB where it must be confessed, the fact of looking at worms limply wiggle is nothing particularly amusing: but it is motivating to feel that one helps science and research, *a fortiori* when the *medical research* aspect is brought forward;

– evolutionary game:

Players aquire experience by playing; they answer in an increasingly more relevant manner to increasingly more specific and more specialized questions. For example, a game of *associated ideas* to the term *cat* is interesting to discover JEUXDEMOTS, but subsequently, players prefer more difficult instructions, otherwise they may get bored. It is necessary that the level of difficulty grows together with the player's experience. It is also possible to define difficulty levels (for example, beginner, normal, confirmed and expert) corresponding to the rarity of the proposed words or the difficulty of the required lexical relations in the case of a lexical game;

– short games:

The shorter the games are, the less the potential player hesitates to play. The number of games played in a session is strictly contravariant with the average duration of the games. This property implies a low granularity of the desired resources to be collected, and therefore rather concerns microtask-based games. In general, lexical approaches give rise to microtasks whose granularity is very fine and whose execution time is very short. In the case of JEUXDEMOTS, a survey has showed that the players who planned to play for 10 min rather spent, on average, 25 min on the game;

– size and portability of the game:

If a game presents a sufficiently simple interface to be playable on smartphones, it is possible to play anywhere and almost at any time (waiting queue, public transports, etc.). Players play more often and it may possibly be a fashion trend caused by word of mouth; it was particularly the case for AKINATOR during its launching on smartphones.

For designers

When the targeted aim is the creation of a quality resource, this acquisition method presents a certain number of advantages:

– the produced resources are the fruit of consensus results, shared by a large number of (or even by a very large number of) players;

– the result is the creation of a database of common general knowledge to which more expert knowledge is necessarily added;

– the cost of resources is very low: it is that of development, then of maintenance and game animation; it should be noted that the last point is not to be neglected, or else a certain disaffection for the game could be quickly observed;

– depending on the number of players and their activity, the acquisition can be relatively fast (for example, in the order of over 1,000 occurrences of lexical relations per day in the context of the JDM project);

– a resource acquisition game must target a segment of the population as wide as possible. However, the very nature of handled objects can greatly reduce this hope. Thus, the skills required to play DIZEEZ (a fine and in-depth knowledge about gene/disease associations) strongly restrict the number of potential players. Conversely, JDM is intended for a much broader and varied audience: 70% of the players are between 30 and 60 years old, 60% are women and almost all of them have a study level equivalent to a Diploma of Higher Education or higher;

– from an ethical point of view, it should be finally observed that the solicitation of contributors using games does not seem to pose as many problems as in a remunerated context [SAG 11], even if at first glance the tasks can seem very similar. In addition, if the game is well designed, the motivation of people to play well is a guarantee of quality of the produced resource, which is far from being assured in the context of a (poorly) remunerated job. Finally, the motivation is very high if the *well-designed* game is at the service of an interesting project and produces free and accessible data.

Rotation of players

The rotation (renewal of players) can be relatively short and all the more so in the presence of a fashion trend within a subgroup or within a small community of players. According to our experience about JDM, the rotation lasts about 3 weeks. We have measured this duration in a continuous manner after each information campaign, either during scientific conferences or among students. A small number of players continue to play for a long time (several years); others discover the game, play some other games and never come back. Are these players the ones who are *unstable* by nature, permanently looking for new experiences, or disappointed by this game? In the latter case, why?

We have also found that a certain number of players come back to the game after a long period of inactivity. Sometimes, it is just to play again, but usually with a different attitude, more focused; for example, they play TOTAKI to verify if the system knows some specific information.

A relatively short rotation of players is a disadvantage: over a short period, there is not enough time, first, to acquire sufficient experience and practice to see any progress and, second, to discover and experience all the subtleties and all the game modalities, and too often the player remains at a basic level. However, it should be indicated that there is a significant proportion of players with an addictive behavior, among whom the system of points and ranking gives rise to a productive emulation: their willingness to play as much as possible and to push the limits drives them to inform themselves, and allows the acquisition of resources that are more akin to expert knowledge.

However, the rotation allows new players, novice by definition, to intervene in the game. The latter perpetuate the supply of naive information (as opposed to expert resources). They also allow the often rapid evolution of ideas to be better taken into account, especially those strongly related to current events, by providing original associations, related to the current context.

Cheating and vandalism

We have denoted two kinds of cheating. In the first place, this allows bypassing certain restrictive rules of the game (for example, the time limitation in JDM): this type of cheating does not question the quality of the produced resources; it may even improve them, since by allowing constraints to be bypassed, it makes things easier. However, it is harmful insofar as, if it becomes obvious that some players cheat, this causes some disgust in others, which may result in a disaffection for the game.

In order to compensate for this disadvantage, we are trying to uncover the *sensible areas* using pilot players, preferably skilled in computing: this kind of cheating is actually caused by players who have enough computing knowledge to exploit the flaws of the game by intervening on the source code. Moreover, numerous attacks have taken place in the context of JDM, notably by injecting computer code. In general, it is the first hours of such a project which constitutes the critical phase for this type of risk.

The second type of cheating is more serious and aims to "rot" the database (i.e. to deliberately cause the insertion of erroneous data) and is more related to vandalism than to the desire of winning by mischievous means. The concern for the designers is to minimize the risk at any cost, starting at the design stage, since this kind of cheating may irremediably jeopardize the quality of the resource which they try to constitute: tedious, even costly to detect and to manually neutralize (by an expert), it is extremely difficult, almost impossible, to detect in an automatic manner. As a result, assuming that the system is able to detect an incongruous information (which is already far from being obvious and which poses the insolvable question of criteria), to systematically classify it as *wrong* and to eliminate it would be counterproductive: *incongruous* does not necessarily mean *wrong*. Furthermore, the deprivation of such original but potentially relevant information is contrary to the desired objectives, which are to enrich the system by providing it with knowledge elements it lacks. If such a system was capable of self-controlling and sorting the data provided, then the project would not make sense.

In many GWAP, data acquisition is based on the principle whereby they are considered valid as much as they have been proposed simultaneously and independently by at least two players. This is why games anonymously and asynchronously confront the responses of two players to an instruction, and the system only stores their common responses (i.e. the intersection of the two sets of data). Thus, if one of the players attempts to sabotage the game by answering nonsense, first, it will not be inserted into the database and, second, the result displayed will discourage malicious players by highlighting the failure of the attempt (none of the ineptitudes is retained). And if sufficient activity is maintained within the game, the probability that two vandal players meet remains low.

Social aspect: the group of players

At a time where social networks have become an unavoidable reality, the community aspect and the feeling of belonging to a group are often largely significant regarding the interest that a game may generate, and GWAP are no exception.

With respect to the durability of a game, the existence of a community of players, in which there are interactions, presents several advantages:

– by playing alone in his/her corner, the player gets quickly bored; the ability to visualize with the interface that other people play simultaneously causes a certain emulation, even if the player does not actually know them;

– providing players with a forum where they can debate, exchange views, approve or contest certain of each-other's suggestions, not only consolidates the group by promoting conviviality, but also makes the players increasingly more expert. Furthermore, any means of communication at their disposal gives them the opportunity to provide feedback about the game and its modalities, which is important for designers;

– there may be a form of appropriation of the game by the players. For example, a player may decide to focus on a theme that is of particular interest to him/her ("I am the expert in this field") and make a point to provide comprehensive and carefully controlled data. This form of specialization can set an example through the interactions within the group and bring other players to also feel they are specialists in a domain. In parallel to the general knowledge, the network will be enhanced by *specialized* knowledge, and this, relating to various fields.

Final thoughts

In this book, we have shown how GWAPs, or games with a purpose, can be employed to easily acquire quality resources at a low cost. However, this assumes that they fulfill a certain number of conditions. Their ludic interest, efficacy and/or the popularity, the media coverage of the project which they support – the four ideally – should not only attract, but also build loyalty in a sufficient number of players so that the generated resource has both relevance and quality. And this is achieved by targeting the widest audience possible while taking account of its diversity, for some the ludic aspect will largely dominate, while others will be seduced by the idea of participating in something useful. *Citizen science* enjoys an increasing success, and many people are willing to engage in this type of project, and even if they do not clearly master the whys and wherefores, but simply because it is rewarding.

For the design of a GWAP, the difficulty of having a ludic cycle covariant with the data acquisition cycle is added to the typical constraints inherent in the design of an Internet game. This means that we must strive to make the interest of the player and the quality of the data it produces interdependent through the mechanism of the game. In other words, a player who plays well (therefore, who produces quality data) should be rewarded by the game in order to encourage this way of playing. On the other hand, the absence of reward, or even the penalty for the player who plays badly, must be paired with the mechanism that avoids bad data to be collected.

The use of GWAP is relatively recent because it is related to the emergence of the Internet. With regard to the acquisition of lexical-semantic data, resorting to GWAP is not in opposition, but appears as a complement to traditional methods, primarily manual and that exist for several centuries. Thus, the lexical network obtained with the JEUXDEMOTS project is not intended to supplant resources such as WordNet or regular dictionaries, but more modestly to be a complement to these reference tools, and to demonstrate the real effectiveness of approaches based on game folksonomy. On the other hand, some other games, mainly macrotask games, do not constitute a complement to conventional methods, but rather an alternative that could allow data acquisition or solving problems whose solution is not known or cannot be envisaged through traditional methods.

In the context of the JEUXDEMOTS project, we have observed that collecting information was easily carried out with games and with a very satisfactory quality, whether the information is related to lexis or to common sense. The authors of GWAP who have published estimates of the quality of resources obtained are overwhelmingly satisfied. However, we have found that if the information requires extensive expertise, the results are much more random. For example, with regard to the creation of a lexical network, only a small amount of data requires linguistic proficiency that exceeds the intuition of the speaker; on the other hand, more complex tasks, such as the creation of a database of syntactic analysis trees or translation alignments, are intended for a much more restricted public insofar as they require much more expertise. Findings are the same for two medical GWAP: DIZEEZ and THE CURE, as described in Chapter 2. Finally, it should be noted that a certain number of resources thus acquired by GWAP can advantageously be evaluated with another GWAP, which must obviously present modalities orthogonal to those of acquisition. One of the major advantages of this kind of approach for the evaluation is that sampling is performed implicitly by the players themselves.

Thus, *the cycle is complete*: the players are the ones who, through a ludic activity, helped to acquire a resource. The players are also the ones who validate the acquired resource, while always *having fun*. There is no doubt that Galileo, seeing the results obtained by this *virtuous cycle*, would reaffirm this time high and strong "*E pur si muove*!".

Bibliography

[AHN 04] VON AHN L., DABBISH L., "Labeling images with a computer game", *Proceedings of the SIGCHI Conference on Human Factors in Computing Systems*, New York, NY, ACM Press, pp. 319–326, 2004.

[AHN 06a] VON AHN L., "Games with a purpose", *Computer*, vol. 39, no. 6, pp. 92–94, 2006.

[AHN 06b] VON AHN L., KEDIA M., BLUM M., "Verbosity: a game for collecting common-sense facts", *Proceedings of ACM CHI 2006 Conference on Human Factors in Computing Systems*, pp. 75–78, 2006.

[AHN 08a] VON AHN L., DABBISH L., "Designing games with a purpose", *Communication of the ACM*, vol. 51, no. 8, pp. 58–67, 2008.

[AHN 08b] VON AHN L., LIU R., BLUM M., "Peekaboom: a game for locating objects in images", *Proceedings of CHI '06*, pp. 55–64, 2008.

[ALO 09] ALONSO O., MIZZARO S., "Can we get rid of trec assessors? using mechanical turk for relevance assessment", *SIGIR '09: Workshop on The Future of IR Evaluation*, Boston, 2009.

[ALO 08] ALONSO O., ROSE D.E., STEWART B., "Crowdsourcing for relevance evaluation", *SIGIR Forum*, vol. 42, no. 2, pp. 9–15, 2008.

[ART 08] ARTSTEIN R., POESIO M., "Inter-coder agreement for computational linguistics", *Computational Linguistics*, vol. 34, no. 4, pp. 555–596, 2008.

[ATT 10] ATTARDI G., GALOAP TEAM, "Phratris – A Phrase Annotation Game", Demo presented at INSEMTIVES Game Idea Challenge, available at https://gate.ac.uk/sale/talks/gate-course-jun14/module-2-socmed/refs.pdf, 2010.

[CAL 09] CALLISON-BURCH C., "Fast, cheap, and creative: Evaluating translation quality using amazon's mechanical turk", *Proceedings of the 2009 Conference on Empirical Methods in Natural Language Processing*, Association for Computational Linguistics, vol. 1, pp. 286–295, 2009.

[CHA 08] CHAMBERLAIN J., POESIO M., KRUSCHWITZ U., "Phrase detectives: a web-based collaborative annotation game", *Proceedings of the International Conference on Semantic Systems (I-Semantics'08)*, Graz, p. 8, 2008.

[CHA 09a] CHAMBERLAIN J., POESIO M., KRUSCHWITZ U., "A new life for a dead parrot: incentive structures in the Phrase Detectives game", *Proceedings of the WWW 2009 Workshop on Web Incentives (WEBCENTIVES'09)*, Madrid, p. 6, 2009.

[CHA 09b] CHAMBERLAIN J., POESIO M., KRUSCHWITZ U., "A demonstration of human computation using the Phrase Detectives annotation game", *HCOMP '09: Proceedings of the ACM SIGKDD Workshop on Human Computation*, pp. 23–24 2009.

[CHA 12] CHAMBERLAIN J., KRUSCHWITZ U., POESIO M., "Motivations for participation in socially networked collective intelligence systems", *Proceedings of Collective Intelligence (CI'12)*, p. 8, available at http://www.ci2012.org/, 2012.

[CHA 13] CHAMBERLAIN J., FORT K., KRUSCHWITZ U., LAFOURCADE M., POESIO M., "Using games to create language ressources: successes and limitations of the approach", *Theory and Applications of Natural Language Processing*, GUREVYCH I., KIM J. (eds), Springer, p. 42, 2013.

[CHK 05a] CHKLOVSKI T., GIL Y., "Improving the design of intelligent acquisition interfaces for collecting world knowledge from web contributors", *Proceedings of K-CAP '05*, pp. 35–42, 2005.

[CHK 05b] CHKLOVSKI T., "Collecting paraphrase corpora from volunteer contributors", *Proceedings of K-CAP '05*, New York, NY, ACM, pp. 115–120, 2005.

[COO 10] COOPER S., KHATIB F., TREUILLE A., BARBERO J., LEE J., BEENEN M., LEAVER-FAY A., BAKER D., POPOVIC Z., FOLDIT PLAYERS., "Predicting protein structures with a multiplayer online game", *Nature*, vol. 466, pp. 756–760, 2010.

[DON 06] DONG Z., DONG Q., "HowNet and the Computation of Meaning", *WorldScientific*, London, p. 316, available at http://www.keenage.com/html/e_index.html, February 2006.

[FAB 06] FABRE C., BOURIGAULT D., "Extraction de relations sémantiques entre noms et verbes au-delà des liens morphologiques", *Proceedings of TALN'06*, Leuven, Belgique, pp. 121–126, 2006.

[FEL 98] FELLBAUM C., *Wordnet: An Electronic Lexical Database*, MIT Press, Cambridge, MA, 1998.

[FEN 09] FENG D., BESANA S., ZAJAC R., "Acquiring high quality non-expert knowledge from on-demand workforce", *Proceedings of the 2009 Workshop on The People's Web Meets NLP: Collaboratively Constructed Semantic Resources, People's Web '09*, Morristown, NJ, Association for Computational Linguistics, pp. 51–56, 2009.

[FIS 09] FISHER L., *The Perfect Swarm: The Science of Complexity in Everyday Life*, Basic Books, 2009.

[FOR 11] FORT K., ADDA G., COHEN K.B., "Amazon mechanical turk: Gold mine or coal mine?", *Computational Linguistics*, vol. 37, no. 2, pp. 413–420, available at http://www.mitpressjournals.org/doi/abs/10.1162/COLI_a_00057#.VVJaUjGsX00, June 2011.

[FOR 14] FORT K., GUILLAUME B., CHASTANT H., "Creating Zombilingo, a game with a purpose for dependency syntax annotation", *Proceedings of the First International Workshop on Gamification for Information Retrieval, GamifIR'14*, p. 5, available at https://hal.inria.fr/hal-00969157/PDF/4-Fort.pdf, 2014.

[GAL 14] GALA N., ZOCK M. (eds), "*Ressources Lexicales – Contenu, construction, utilisation, évaluation*", John Benjamins Publishing Company, 2013.

[GOO 13] GOOD B.M., SU A.I., "Crowdsourcing for Bioinformatics", *Bioinformatics*, vol. 29, no. 16, pp. 1925–1933, 2013.

[GOO 14] GOOD B.M., LOGUERCIO S., GRIFFITH O.L., NANIS M., WU C., SU A.I., "*The Cure: Design and Evaluation of a Crowdsourcing Game for Gene Selection for Breast Cancer Survival Prediction*", JMIR Serious Games, vol. 2, no. 2, p. e7, 2014.

[GRE 10] GREEN N., BREIMYER P., KUMAR V., SAMATOVA N.F., "PackPlay: mining semantic data in collaborative games", *Proceedings of the 4th Linguistics Annotation Workshop, LAW IV'10*, 2010.

[HAN 10] HAND E., "Citizen science: people power", *Nature*, vol. 466, pp. 685–687, 2010.

[HLA 09] HLADKA B., MIROVSKY J., SCHLESINGER P., "Play the language: play coreference", *Proceedings of the ACL-IJCNLP 2009 Conference Short Papers*, Association for Computational Linguistics, pp. 209–212, 2009.

[HLA 11] HLADKA B., MIROVSKY J., KOHOUT J., "An attractive game with thedocument:(im) possible?", *The Prague Bulletin of Mathematical Linguistics*, no. 96, pp. 5–26, available at https://ufal.mff.cuni.cz/pbml/96/art-hladka-mirovsky-kohout.pdf, October 2011.

[JOH 98] JOHNSON N.L., RASMUSSEN S., JOSLYN, C., ROCHA L., SMITH S., KANTOR M., "Symbiotic intelligence: self-organizing knowledge on distributed networks driven by human interaction", *Proceedings of the Sixth International Conference on Artificial Life*, MIT Press, pp. 404–406, available at https://books.google.fr/, 1998.

[JOU 11] JOUBERT A., LAFOURCADE M., SCHWAB D., ZOCK M., "Évaluation et consolidation d'un réseau lexical grâce à un assistant ludique pour le "mot sur le bout de la langue"", *Proceedings of TALN'11*, Montpellier, France, pp. 295–306, 2011.

[JOU 12] JOUBERT A., LAFOURCADE M., "A new dynamic approach for lexical networks evaluation", *Proceedings of the 8th International Conference on Language Resources and Evaluation (LREC'12)*, Istanbul, Turkey, 23–25 May, 2012.

[JUR 08] JURAFSKY D., MARTIN J.H., *Speech and Language Processing: An Introduction to Natural Language Processing, Computational Linguistics, and Speech Recognition*, 2nd edition, Prentice-Hall, Computers, p. 988, available at http://books.google.fr/books/about/Speech_and_Language_Processing.html?id=fZmj5UNK8AQC&redir_esc=y, 2009.

[JUR 14] JURGENS D., NAVIGLI R., "It's all fun and games until someone annotates: video games with a purpose for linguistic annotation", *Transactions of the Association for Computational Linguistics (TACL)*, vol. 2, pp. 449–464, 2014.

[KAN 01] KANEFSKY B., BARLOW N., GULICK V., "Can distributed volunteers accomplish massive data analysis tasks?", *Lunar and Planetary Science*, vol. 32, p. 2, 2001.

[KAW 12] KAWRYKOW A., ROUMANIS G., KAM A., KWAK D., LEUNG C., WU C., ZAROUR E., PLAYERS P., SARMENTA L., BLANCHETTE M., WALDISPÜHL J., "Phylo: a citizen science approach for improving multiple sequence alignment", *PLoS ONE*, vol. 7, no. 3, p. e31362, available at http://ti.arc.nasa.gov/m/pub-archive/212h/0212%20%28Kanefsky%29.pdf, 2001.

[KAZ 09] KAZAI G., MILIC-FRAYLING N., COSTELLO J., "Towards methods for the collective gathering and quality control of relevance assessments", *Proceedings of the 32nd international ACM SIGIR conference on Research and development in information retrieval, SIGIR '09*, New York, NY, ACM, pp. 452–459, 2009.

[KAZ 11] KAZAI G., "In search of quality in crowdsourcing for search engine evaluation", *Proceedings of the 33rd European Conference on Information Retrieval (ECIR'11)*, Lecture Notes in Computer Science, Springer, vol. 6611, pp. 165–176, 2011.

[KHA 11] KHATIB F., COOPER S., TYKA M.D., XU K., MAKEDON I., POPOVIC Z., BAKER D., FOLDIT PLAYERS, "Algorithm discovery by protein folding game players", *Proceedings of the National Academy of Sciences*, vol. 108, no. 47, pp. 18949–18953, 2011.

[KIM 14] KIM J.S., GREENE M.J., ZLATESKI A., LEE K., RICHARDSON M., TURAGE S.C., PURCARO M., BALKAM M., ROBINSON A., BEHABADI B.F., CAMPOS M., DENK W., SEUNG H.S.,EYEWIRERS., "Space-time wiring specificity supports direction selectivity in the retina", *Nature*, vol. 509, pp. 331–336, 2014.

[KOL 10] KOLLER A., STRIEGNITZ K., GARGETT A., BYRON D., CASSELL J., DALE R., MOORE J., OBERLANDER J., "Report on the second nlg challenge on generating instructions invirtual environments (give-2)", *Proceedings of the 6th INLG*, Dublin, p. 8, 2010, available at http://www.aclweb.org/anthology/W10-4233.

[KOS 05] KOSTER R., "A Theory of Fun for Game Design", *Paraglyph*, p. 244, available at http://www.goodreads.com/book/show/18182.A_Theory_of_Fun_for_Game_Design, 2005.

[KUO 09] KUO Y.L., LEE J.C., CHIANG K.Y., WANG R., SHEN E., CHAN C.W., HSU Y-J.J., "Community-based game design: experiments on social games for commonsense data collection", *Proceedings of the ACM SIGKDD Workshop on Human Computation, HCOMP'09*, New York, NY, pp. 15–22, 2009.

[KWA 13] KWAK D., KAM A., BECERRA D., ZHOU Q., HOPS A, ZAROUR E., KAM A., SARMENTA L., BLANCHETTE M., WALDISPÜHL J., "Open-Phylo: a customizable crowd-computing platform for multiple sequence alignment", *Genome Biology*, vol. 14, p. R116, 2013.

[LAF 07] LAFOURCADE M., "Making people play for lexical acquisition", *Proceedings of SNLP 2007, 7th Symposium on Natural Language Processing, Pattaya*, Thailand, p. 8, 2007.

[LAF 09] LAFOURCADE M., ZAMPA V., "PtiClic: a game for vocabulary assessment combining JeuxDeMots and LSA", *In Proceedings of CICLing (Conference on Intelligent text processing and Comptational Linguistics)*, Mexico, 1–7 mars, p. 10, 2009.

[LAF 10] LAFOURCADE M., JOUBERT A., "Computing trees of named word usages from a crowdsourced lexical network", *Investigationes Linguisticae*, vol. 21, pp. 39–56, 2010.

[LAF 14a] LAFOURCADE M., LE BRUN N., ZAMPA V., "Crowdsourcing word-color associations", *19th International Conference on Application of Natural Language to Information Systems (NLDB 2014)*, 18–20 June 2014, Montpellier, France, p. 6, 2014.

[LAF 14b] LAFOURCADE M., FORT K., "Propa-L: a semantic filtering service from a lexical network created using Games With A Purpose", *Proceedings of the Ninth International Conference on Language Resources and Evaluation (LREC)*, Reykjavik, Iceland, pp. 1676–1681, 26–31 May 2014.

[LEE 14] LEE J., KLADWANG W., LEE M., CANTU D., AZIZYAN M., KIM H., LIMPAECHER A., YOON S., TREUILLE A., DAS R., ETERNA PARTICIPANTS., "RNA design rules from a massive open laboratory", *Proceedings of the National Academy of Sciences*, vol. 111, no. 6, pp. 2122–2127, 2014.

[LEN 95] LENAT D., "CYC: a large-scale investment in knowledge infrastructure", *Communications of the ACM*, vol. 38, no. 11, pp. 33–38, 1995.

[LIE 07] LIEBERMAN H., SMITH D., TEETERS A., "Common consensus: a web-based game for collecting commonsense goals", Workshop on Common Sense for Intelligent Interfaces, *ACM International Conference on Intelligent User Interfaces (IUI-07)*, Honolulu, Hawaii, p. 6, available at http://web.media.mit.edu/~lieber/Publications/Common-Consensus.pdf, 2007.

[LUE 12] LUENGO-OROZ M.A., ARRANZ A., FREAN J., "Crowdsourcing malaria parasite quantification: an online game for analyzing images of infected thick blood smears", *JMIR*, vol. 14, no. 6, p. e167, 2012.

[MAL 09] MALONE T.W., LAUBACHER R., DELLAROCAS C., Harnessing crowds: mapping the genome of collective intelligence, pp. 327–335, available at SSRN: http://ssrn.com/abstract=1381502, 3 February 2009.

[MAR 93] MARCUS M., SANTORINI B., MARCINKIEWICZ M.A., "Building a large annotated corpus of english: the penn treebank", *Computational Linguistics*, vol. 19, no. 2, pp. 313–330, 1993.

[MAR 12] MARSHALL J., "Victory for crowdsourced biomolecule design", *Nature News*, 1038/nature.2012.9872, 2012.

[MAS 10] MASON W., WATTS D.J., "Financial incentives and the "performance of crowds"", *ACM SIGKDD Explorations Newsletter archive*, vol. 11, no. 2, pp. 100–108, available at http://dl.acm.org/citation.cfm?id=1809422, December 2010.

[MAV 12] MAVANDADI S., DIMITROV S., FENG S., YU F., SIKORA U., YAGLIDERE O., PADMANABHAN S., NIELSEN K.. OCAN A., "Distributed medical image analysis and diagnosis through crowd-sourced games: a malaria case study", *PLoS ONE*, vol. 7, no. 5, p. e37245, 2012.

[MIC 06] MICHAEL D.R., CHEN S., *Serious Games: Games that Educate, Train and Inform*, Thomson Course Technology, p. 287, 2006.

[MIH 03] MIHALCEA R., CHKLOVSKI T., "Open MindWord Expert: creating large annotated data collections with web users help", *Proceedings of the EACL 2003*, Workshop on Linguistically Annotated Corpora (LINC 2003), pp. 53–60, available at http://www.aclweb.org/anthology/W03-2408, 2003.

[MIL 90] MILLER, G.A., BECKWITH, R., FELLBAUM, C., GROSS, D., MILLER, K.J., "Introduction to WordNet: an on-line lexical database", *International Journal of Lexicography*, vol. 3, no. 4, pp. 235–244, 1990.

[MIL 95] MILLER G., "WordNet: a lexical database for english", *Communications of the ACM*, vol. 38, no. 11, pp. 39–41, 1995.

[MRO 08] MROZINSKI J., WHITTAKER E., FURUI S., "Collecting a whyquestion corpus for development and evaluation of an automatic QA-system", *Proceedings of ACL-08: HLT*, Columbus, Ohio, pp. 443–451, 2008.

[NAV 10] NAVIGLI R., PONZETTO S., "BabelNet: building a very large multilingual semantic network", *Proceedings of the 48th Annual Meeting of the Association for Computational Linguistics*, Uppsala, Sweden, pp. 216–225, 2010.

[PLO 98] PLOUX S., VICTORRI B., "Construction d'espaces sémantiques à l'aide de dictionnaires de synonymes", *Proceedings of Traitement Automatique des Langues*, vol. 39/1, pp. 161–182, 1998.

[POE 04] POESIO M., "Discourse annotation and semantic annotation in the GNOME corpus", *Proceeding DiscAnnotation '04 Proceedings of the 2004 ACL Workshop on Discourse Annotation*, pp. 72–79, available at http://dl.acm.org/ citation.cfm?id=1608948, 2004.

[POE 06] POESIO M., STURT P., ARSTEIN R., FILIK R., "Underspecification and anaphora: Theoretical issues and preliminary evidence", *Discourse Processes*, vol. 42, no. 2, pp. 157–175, 2006.

[PUS 93] PUSTEJOVSKY J., "The generative lexicon", *Computational Linguistics*, pp. 409–441, 1993.

[QUI 11] QUINN A., BEDERSON B., "Human computation: a survey and taxonomy of a growing field",*Proceedings of the SIGCHI Conference on Human Factors in Computing Systems*, pp. 1403–1412, available at http://dl.acm.org/ citation.cfm?id=1979148&dl=ACM&coll=DL&CFID=511124876&CFTOKEN=94270210 and http://www.chi2011.org/, 2011.

[RAF 09] RAFELSBERGER W., SCHARL A., "Games with a purpose for social networking platforms", *Proceedings of the 20th ACM conference on Hypertext and hypermedia*, ACM, pp. 193–198, 2009..

[SAG 08] SAGOT B., FISER D., "Construction d'un wordnet libre du français à partir de ressources multilingues", *Proceedings of TALN 2008*, Avignon, France, p. 10, available at http://alpage.inria.fr/∼sagot/pub/TALN08.pdf, 2008.

[SAG 11] SAGOT B., FORT K., ADDA G., MARIANI J., LANG B., "Un turc mécanique pour les ressources linguistiques: critique de la myriadisation du travail parcellisé", *Traitement Automatique des Langues Naturelles (TALN 2011)*, Montpellier, France, 27th June–1st July 2011.

[SCH 13] SCHROPE M., "Solving thought problems with games", *Proceedings of the National Academy of Sciences*, vol. 110, no. 18, pp. 7104–7106, 2014.

[SIO 08] SIORPAES K., HEPP M., "Games with a purpose for the Semantic Web", *IEEE Intelligent Systems*, vol. 23, no. 3, pp. 50–60, 2008.

[SMA 09] SMADJA F., "Mixing financial, social and fun incentives for social voting", *World Wide Web Internet And Web Information Systems*, p. 6, available at http://fr.slideshare.net/MbayeSmadja/webcentives09-smadja, April 2009.

[SNO 08] SNOW R., O'CONNOR B., JURAFSKY D., NG A.Y., "Cheap and fast – but is it good?: evaluating non-expert annotations for natural language tasks", *EMNLP'08: Proceedings of the Conference on Empirical Methods in Natural Language Processing*, Morristown, NJ, Association for Computational Linguistics, pp. 254–263, 2008.

[STE 11] STEINMAYR B., WIESER C., KNEISSL F., ET BRY F., "Karido: a GWAP for telling artworks apart", *16th International Conference on Computer Games*, pp. 193–200, 27–30 July 2011.

[STR 12] STRAUSS S., "Gamers outdo computers at matching up disease genes – Computer game crowdsources DNA sequence alignments across different species", *Nature News*, 1038/nature.2012.10203, 2012.

[STÖ 12] STÖRKLE F., "Combino – a GWAP for generating combined tags", Rapport de stage de MASTER, Institut d'Informatique – LMU – Munich, p. 53, 2012.

[TCH 14] TCHERNIA R., "Science participative, les joueurs aux manettes", *Le Monde*, 5 March 2014.

[THA 11] THALER S., SIORPAES K., SIMPERL E., HOFER C., A survey on games for knowledge acquisition, STI Technical Report, 2011.

[VAN 14] VANNELLA D., JURGENS D., SCARFINI D., TOSCANI D., NAVIGLI R.,"Validating and extending semantic knowledge bases using video games with a purpose", *Proceedings of the Annual Meeting of the Association for Computational Linguistics, ACL'14*, p. 11, available at http://wwwusers.di.uniroma1.it/~navigli/pubs/ACL_2014_Vannellaetal.pdf, 2014.

[VEN 13] VENHUIZEN N., BASILE V., EVANG K., BOS J., "Gamification for word sense labeling", *Proceedings of the 10th International Conference on Computational Semantics (IWCS 2013) – Short Papers*, Potsdam, Germany, pp. 397–403, 2013.

[VÉR 90] VÉRONIS J., IDE N., "Word sense disambiguation with very large neural networks extracted from machine readable dictionaries", *Proceedings of 13th International Conference on Computational Linguistics (COLING'90)*, Helsinki, vol. 2, pp. 389–394, 1990.

[VIC 08] VICKREY D., BRONZAN A., CHOI W., KUMAR A., TURNER-MAIER J., WANG A., KOLLER D., "Online word games for semantic data collection", *Proceedings of the Conference on Empirical Methods in Natural Language Processing, EMNLP'08*, Honolulu, pp. 533–542, October, 2008.

[VOS 98] VOSSEN P., "*EuroWordNet: A Multilingual Database with Lexical Semantic Networks*", Kluwer Academic Publishers, Dordrecht, The Netherlands, p. 179, available at http://journals.cambridge.org/action/displayAbstract?fromPage=online&aid=188411&fileId=S1351324903223299, 1998.

[WIE 13] WIESER C., BRY F., BÉRARD A., LAGRANGE R., "ARTigo: building an artwork search engine with games and higher-order latent semantic analysis", *Proceedings of Disco 2013, Workshop on Human Computation and Machine Learning in Games at the International Conference on Human Computation (HComp)*, Palm Springs, California, USA, p. 6, available at http://www.en.pms. ifi.lmu.de/publications/PMS-FB/PMS-FB-2013-3/PMS-FB-2013-3-paper.pdf, 6–9th November 2013.

[ZAM 10] ZAMPA V., LAFOURCADE M., "PtiClic et PtiClic-kids: jeux avec les mots permettant une double acquisition", *Proceedings of TICE2010*, 7th Seminar TICE, Nancy, p. 10, 2010.

[ZES 09] ZESCH T., GUREVYCH I., "*Wisdom of Crowds Versus Wisdom of Linguists Measuring the Semantic Relatedness of Words*", Natural Language Engineering, Cambridge University Press, pp. 25–59, 2009.

Index

Other titles from

iSTE

in

Cognitive Science and Knowledge Management

2014

DELPECH Estelle Maryline
Comparable Corpora and Computer-assisted Translation

FARINAS DEL CERRO Luis, INOUE Katsumi
Logical Modeling of Biological Systems

MACHADO Carolina, DAVIM J. Paulo
Transfer and Management of Knowledge

SAAD Inès, ROSENTHAL-SABROUX Camille, GARGOURI Faiez
Information Systems for Knowledge Management

TORRES-MORENO Juan-Manuel
Automatic Text Summarization

2013

TURENNE Nicolas
Knowledge Needs and Information Extraction: Towards an Artificial Consciousness

ZARATÉ Pascale
Tools for Collaborative Decision-Making

2011

DAVID Amos
Competitive Intelligence and Decision Problems

LÉVY Pierre
The Semantic Sphere: Computation, Cognition and Information Economy

LIGOZAT Gérard
Qualitative Spatial and Temporal Reasoning

PELACHAUD Catherine
Emotion-oriented Systems

QUONIAM Luc
Competitive Intelligence 2.0: Organization, Innovation and Territory

2010

ALBALATE Amparo, MINKER Wolfgang
Semi-Supervised and Unsupervised Machine Learning: Novel Strategies

BROSSAUD Claire, REBER Bernard
Digital Cognitive Technologies

2009

BOUYSSOU Denis, DUBOIS Didier, PIRLOT Marc, PRADE Henri
Decision-making Process

MARCHAL Alain
From Speech Physiology to Linguistic Phonetics

PRALET Cédric, SCHIEX Thomas, VERFAILLIE Gérard
Sequential Decision-Making Problems / Representation and Solution

SZÜCSAndras, TAIT Alan, VIDAL Martine, BERNATH Ulrich
Distance and E-learning in Transition

2008

MARIANI Joseph
Spoken Language Processing

Printed and bound by CPI Group (UK) Ltd, Croydon, CR0 4YY
13/09/2026
14961017-0008